MERCURY

Kosmos

A series exploring our expanding knowledge
of the cosmos through science and technology
and investigating historical, contemporary
and future developments as well as providing
guidance for all those interested in astronomy.

Series Editor: Peter Morris

Already published:

Jupiter William Sheehan and Thomas Hockey
Mercury William Sheehan
The Moon Bill Leatherbarrow
The Sun Leon Golub and Jay M. Pasachoff

Mercury

William Sheehan

REAKTION BOOKS

To Clark Chapman and Dale Cruikshank, who largely inspired my love of Mercury

Published by Reaktion Books Ltd
Unit 32, Waterside
44–48 Wharf Road
London N1 7UX, UK
www.reaktionbooks.co.uk

First published 2018

Printed and bound in China

A catalogue record for this book is available from the British Library

ISBN 978 1 7891 4012 5

CONTENTS

Preface 7

1 The Scintillating One 9

2 Motions of Mercury 13

3 Through the Telescope 29

4 Rotation 41

5 Mercury Up Close 86

6 Vulcan 131

appendix i: Glossary 153

appendix ii: Basic Data 157

appendix iii: Craters 160

References 165
Further Reading 173
Acknowledgements 174
Photo Acknowledgements 176
Index 178

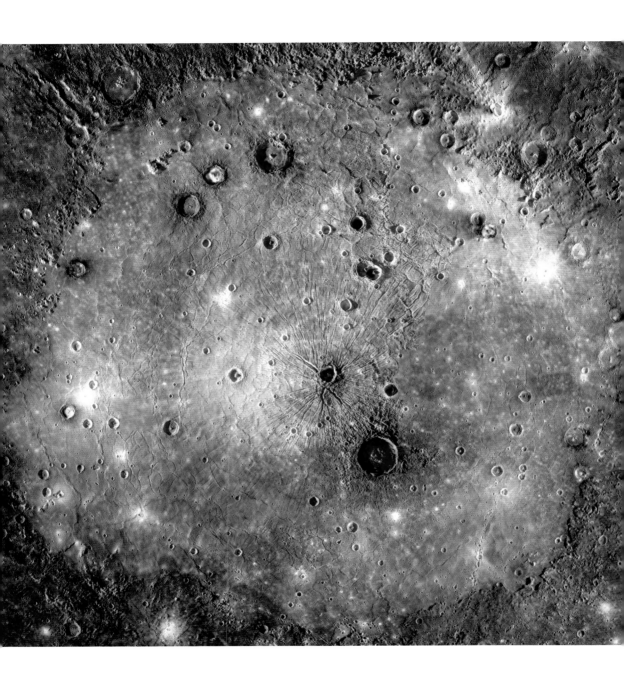

Preface

Mercury ranks among the five planets known in ancient times, and has long had a reputation for being difficult to see – though there is no truth to the legend that even Copernicus could never distinguish it. In fact, it can become quite as bright as Jupiter, and stands out readily when at its greatest elongations from the Sun. From northern latitudes, the best opportunities for seeing Mercury in the evening sky always occur in the spring, and for seeing it in the morning sky in the autumn.

To the naked eye, Mercury appears to twinkle, owing both to its small size (its diameter is only half again larger than the Moon's) and to its being seen through the full thickness of the Earth's roiling atmosphere. Without taking special precautions, a telescope will usually show little more than the moon-like phases characteristic of the inferior planets – those travelling closer to the Sun than the Earth. Historically, a breakthrough in the study of Mercury occurred in the 1880s, when the great Italian astronomer Giovanni Schiaparelli showed the feasibility of observing it in broad daylight. He produced a map of Mercury, and also believed that he had worked out the rotation period, which he put equal to the period of revolution around the Sun – 88 days.

Though other observers – most notably the skilful Graeco-French astronomer Eugène Michel Antoniadi – followed in his footsteps, in fact much of what we thought we knew about Mercury before the

Caloris basin. A colour-enhanced composite overlain on a monochrome mosaic of MESSENGER images. The basin has been flooded by lavas that are colour-coded orange in this mosaic. Post-flooding craters have excavated material from below the surface, which – in the case of larger craters – has exposed low-reflectance material (blue). This likely represents the original basin floor material. Analysis of these craters yields an estimate of the thickness of the volcanic layer: 2.5–3.5 km.

1960s, including the rotation period, was wrong. This says nothing about the skill or dedication of the observers who tried to study it with telescopes from Earth. They made mistakes, but they were hardly blameworthy; rather, they stand as testimony to the difficulty of observing, under challenging conditions, the surface of a world whose markings are less clearly differentiated than those, say, of the Moon or Mars.

Only in the 1960s, when Mercury came within reach of radio telescopes, did our modern understanding of the planet begin, with the discovery of the 58.65-day rotation period. This period is exactly two-thirds of the period of revolution around the Sun, and represents a case of spin-orbit coupling which is unique in the Solar System. Another radio discovery, equally unexpected, is of the existence of bright patches of water ice in permanently shadowed craters near the poles.

Detailed knowledge about Mercury, of course, awaited the spacecraft era. A great step forward was provided by Mariner 10, which made three flybys of the planet in 1974–5. Afterwards there was a long hiatus. It finally ended with the successful mission of the Mercury Surface, Space Environment, Geochemistry and Ranging mission (MESSENGER), which orbited Mercury between 2011 and 2015. Another spacecraft, BepiColombo, a joint project of the European Space Agency and Japan, is set to be launched in 2018, with arrival at Mercury due in 2025.

Though we now know a great deal about Mercury, mysteries remain: not least is the fact that its core has the highest iron content of any body in the Solar System. Though it shares its name with the element also known as quicksilver, it is, in fact, an iron planet. Doubtless the explanation for this distinctive feature will be found in the circumstances of its origin and early evolution as a world.

ONE

THE SCINTILLATING ONE

Elusive in the glow of twilight, Mercury never strays more than 28 degrees from the Sun, so that whenever it ventures to appear it is only as an elusive intruder into our skies that never skims far above the rooftops or hedges. It rises at most two hours before sunrise and sets two hours after sunset, so that it must be seen through the full thickness of the Earth's atmosphere. At such times the small size of its planetary disc and the tumult of the air combine to set it to twinkling. Noticing this, the ancient Greeks referred to it as Stilbon – the 'scintillating one'.

It used to be claimed that Nicolaus Copernicus (1473–1543) never saw it during his lifetime, owing to mists rising from the Frisches Haff near Frauenberg (now Frombork, Poland), in whose cathedral he was a canon. But the story is baseless – after all, as a young man Copernicus spent many years in the favourable climate of Italy, where conditions for viewing Mercury are better than in Poland. In England, conditions are more like those in Poland than Italy; thus a seventeenth-century writer, Goud, claimed that Mercury was an impious lackey of the Sun who rarely showed his head in England, rather like one on the run from his debtors.

In contrast to the Copernicus legend is the story of Gallet, a seventeenth-century cleric at Avignon, who was nicknamed the 'Hermophile' because he managed to see the planet with the naked eye more than a hundred times during his lifetime. Gallet's record

Mercury, the swift-of-foot messenger god, seen here in Glasgow, Scotland.

may have been good for conditions in southern France, but most alert amateurs will have easily exceeded it in a few years of observation.

The trick to seeing Mercury is looking at the right times and having a reasonably unobstructed *tour d'horizon*. It is more often seen, no doubt, than recognized. Despite its supposed difficulty of detection, Mercury ranks among the planets known since earliest times. The fact that it was recognized by many of the star-watching peoples of antiquity is testified by the succession of names they gave to it. The ancient Egyptians, who apparently recognized that Mercury, like Venus, moves round the Sun, called it Sobkou. It was also known in the Nile Valley by a name equivalent to that of the Greek Apollo. The Sumerians called it Bi-ib-bou, the Assyrians and Chaldeans named it Goudoûd, while the Babylonians knew it as Sekhès, and Ninob, Nabou or Nebo. According to Eugène Michel Antoniadi, the last three signified that it was possessed of great and exceptional intelligence. Indeed, Nabou seems to have been the ruler of the universe, since he alone could raise the Sun from its bed.[1]

Venus, Mercury and the crescent Moon make a triangle in the sky. Mercury is the lowest, dimmest 'star', barely visible as a tiny dot to the left of the dark cloud.

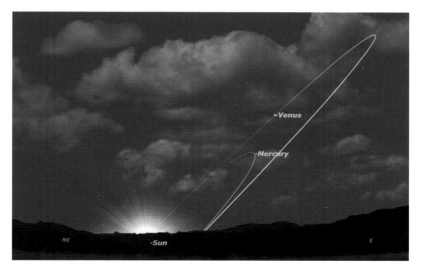

The orbits of Mercury and Venus, the two planets closer to the Sun than the Earth, shown with respect to our horizon.

Among the Greeks, it was not only Stilbon, the scintillating one, but Hermes, the messenger of the gods of Mount Olympus, who announced the rising of the God of Day. (He is one and the same with the Roman Mercury.)

The observer who succeeds in catching sight of it soon becomes aware of the aptness of these appellatives, for Mercury is a world of

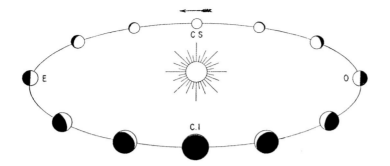

An inferior planet (for example, Mercury) and its positions with respect to the Sun and Earth. Note that maximum elongations bring the planet furthest away from the Sun in our sky.

rapid movement. The favourable opportunities for catching it occur for a week or two round the greatest elongations east or west of the Sun, which occur an average of six times each year. These elongations are not equally favourable, and because of the inclination of the ecliptic, for northern observers the best times to see the flushed little planet as an evening star occur in the spring, and as a morning star in the autumn, as noted earlier. Its movement from night to night is very evident as it quickly loops outwards from and back towards the Sun. If seen as an evening star, it will not return as an evening star for another 115.9 days (its so-called synodic period).

TWO

MOTIONS OF MERCURY

The ancient Greeks held that the Earth was the centre of the universe, and that the Sun and other planets revolved around it. (An exception was Aristarchus of Samos, who in the third century BCE advanced a heliocentric theory in which the Earth was an ordinary planet revolving around the Sun; there were, however, few takers at the time.)

The most elaborate form of the geocentric theory was put forward by Claudius Ptolemy, a Greek astronomer who lived in Alexandria, Egypt, site of the famous library, in the second century CE. His greatest astronomical work is the ἡ μεγίστη (Syntaxis – later known as the *Great Syntaxis* to distinguish it from a lesser collection of astronomical writings; the title was later rendered by the Arabs as *al-majisti*, the *Greatest*, and translated from Arabic to Latin as the *Almagest*, by which it is generally known). Next only to Euclid's *Elements*, it has the distinction of having been the scientific text longest in use, and for over a thousand years remained the supreme authority on astronomy wherever Greek culture and learning survived.

Ptolemy was a synthesizer who perfected the work of his great predecessors, such as Hipparchus of Nicaea (c. 190–c. 120 BCE) and Apollonius of Perga (c. 247–205 BCE), and so successfully that the technical details of their works are largely lost. In order to represent the apparent motions of the planets on the basis of an Earth-centred scheme, they devised a system in which a planet moves in a little circle (known as the epicycle), which in turn moves about a larger circle (the deferent) centred on the Earth. Ptolemy brought this system of epicycles and deferents to its fullest elaboration – thus it is usually referred to as the Ptolemaic system, even though Ptolemy did not, in fact, invent it. In its final form, it was admittedly rather complicated and artificial looking, but it represented the motions of the planets to within the limits of accuracy of the observations at the time, and in those terms it deserves to be judged. The Harvard historian Owen Gingerich has said:

> It is difficult to convey the elegance of Ptolemy's achievement to anyone who has not examined its details. Basically, for the first time in history (so far as we know) an astronomer has shown how to convert specific numerical data into the parameters of planetary models, and from the models has constructed a set of tables that employ some admirably clever mathematical simplifications, and from which solar, lunar, and planetary positions and eclipses can be calculated as a function of any given time.[1]

Ptolemy deserves special plaudits for his theory for the motion of Mars. Mars's motions are so complicated that they earned it a reputation (in the first-century CE Roman natural history writer Pliny's words) as the 'untrackable star'. Ptolemy inherited from his predecessors the basic epicycle construction (by the same token that Bach inherited the fugue and Beethoven the sonata). It was the way he applied it that was ingenious.

As the epicycle turned, a point on its rim followed a looped path swinging in towards the centre of the larger circle, before moving outwards again in reverse. This rather complicated mechanism was required to explain the so-called retrograde motions of Mars and the other outer planets – around the time the planet appears opposite

Ptolemy's epicycle-and-deferent model for the observed motion of planets with respect to the Earth. Note that the Earth is stationary in this model.

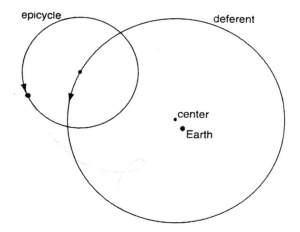

the Sun in the sky, it seems to move through a backward loop, before resuming its direct motion. As we now know, because Copernicus and Kepler showed us, the retrograde movement is merely a reflection – in the outer planet's apparent path – of the Earth's own orbit around the Sun.

That orbit is actually an ellipse of rather high eccentricity, which makes the speed of Mars's motion variable. An observable result is that it sometimes moves through its retrograde loops much faster than at others. Ptolemy managed to explain this (within the limits of accuracy of naked-eye observations) by putting his deferent circle slightly off-centre from the Earth, and adjusting it until he had the speed of Mars's motion just right. Then – and here his tinkering touched on genius – he made it move uniformly not around the centre of the deferent but around another point introduced by Ptolemy in the second century BCE called the equant, so-called because it lay an equal distance on the opposite side of the centre from the Earth (this solution was referred to as the 'bisection of the eccentricity').

Gratified by his success with Mars, Ptolemy did not hesitate to apply it to the rest of the planets. He was successful – except in one bedevilling case: Mercury. Here it was the uncertainty of the observations that created difficulties for him. As we now know, Ptolemy would have done perfectly well had he represented Mercury's motion with the same basic construction he devised for the other planets, but because it is almost impossible to observe in certain positions in its orbit from mid-northern latitudes – particularly at its morning elongations in April and its evening elongations in October – Ptolemy incorrectly deduced that there were two points at which the apparent diameter of Mercury's epicycle would be greatest as seen from the Earth. To accommodate this assumption, he constructed a hypothesis in which the planet could make two close sweeps to the Earth. It worked just like a crank – the centre of the deferent moving about a little circle and thrusting the epicycle back and forth along a line. One cannot help admiring the Alexandrian astronomer's ingenuity. Obviously this little planet, which even at its best just managed to skirt outside the Sun's rays, was destined to give astronomers headaches.

Sun-and-planet gear mechanism in the Whitbread Engine in the Powerhouse Museum, Sydney, Australia, built in 1785. The gear converts the vertical motion of a steam-engine-driven beam into circular motion using a 'planet', a cogwheel placed at the end of the connecting rod of the engine.

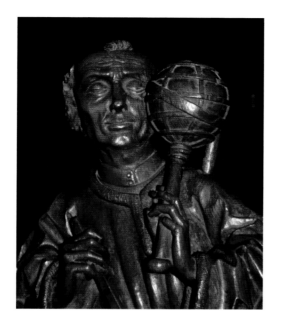

Claudius Ptolemy as imagined by the 15th-century sculptor Jörg Syrlin the Elder via his figure on the choir stalls in Ulm Cathedral.

Here at least Ptolemy literally deserved the epithet allegedly hurled at his system by King Alfonso X of Castile, who sponsored new tables of the planetary motions based on Ptolemy that were to remain standard for two centuries after his death in 1284 CE. Ptolemy's astronomy was, said Alfonso, 'a crank machine . . . it was a pity the Creator did not take advice'.

Parenthetically, the crank mechanism Ptolemy devised for Mercury would reappear at the end of the eighteenth century when the English inventor James Watt borrowed what he called the 'Sun-and-planet gear' in his designs for the steam engine. So perhaps Ptolemy should be regarded in some sense as an unheralded forerunner of the Industrial Revolution.

There is an evocative wooden sculpture of Ptolemy in the Cathedral of Ulm: eyes closed, a faint smile on his face, he holds an armillary sphere in one hand while the other is placed on an Astronomical Radius, or cross-staff. It is an image of smugness and complacency – the face of an astronomer so satisfied with the completeness of his theory that he no longer bothers to look into the heavens. But it is a caricature, and we can be sure that the real Ptolemy did not remotely resemble this image. Instead, he must have been one of the most passionate seekers after celestial truth that has ever lived, and he himself may have written the inspired passage found in some of the old manuscripts of the *Almagest*:

> I know that I am mortal by nature, the creature of a day;
> But when I trace at my pleasure the windings to and fro of the
> heavenly bodies

I no longer touch the earth with my feet:
I stand in the presence of Zeus himself
And take my fill of ambrosia, the food of the gods.

Still Baffling

Ptolemy's last recorded observation was made in 151 CE, and
he must have lived on to at least 170. His Earth-centred system
continued to dominate astronomical thought until 1543, when
Nicolaus Copernicus, a Roman Catholic canon in the Cathedral
of Frauenberg, published his immortal book *De revolutionibus orbium
coelestium* (On the Revolutions of the Celestial Spheres). (It is said
that he received the first printed copies on his deathbed.) Here,
seventeen centuries after Aristarchus, he revived the heliocentric,
or Sun-centred, model – a decisive step. In some ways, however,
Copernicus was a conservative; for instance, he retained the whole
machinery of epicycles, though in fairness by putting the Sun rather
than the Earth at the centre of his system he was able to strike the
main epicycle from the theory of each planet, a considerable
simplification. No less than Ptolemy had done, Copernicus
struggled with Mercury. He placed too much faith in Ptolemy's
observations. (Indeed, he published no new observations of Mercury
of his own.) With evident frustration, he wrote that 'this planet
has . . . inflicted many perplexities and labors on us in our investigation
of its wanderings.'[2]

  It was up to the great work of the Danish astronomer Tycho
Brahe (1546–1601), born three years after Copernicus's death, to
repair the deficiency in astronomical data so that the theoreticians
could advance. For twenty years he owned a magnificent observatory,
Uraniborg, on the island of Hven, in the Sound between Copenhagen
and Elsinore Castle. He had the most accurate instruments money
could buy, and used them with such obsessive care that he was able
to obtain a treasure trove of planetary positions to an accuracy of a

Detail of a 19th-century copy of an earlier painting by an unknown artist of Nicolaus Copernicus.

few minutes of arc – the limit of what was obtainable with the naked eye. (The telescope did not first appear until after his death.) In 1600 he was joined by a gifted young German mathematician, Johannes Kepler (1571–1630), who, following Tycho's death a year later, managed to overcome the resistance of Tycho's heirs and gain control of the master's observing logbooks. By concentrating on Mars, the planet to which Tycho had paid especially close attention, Kepler succeeded after many trials to work out the true shape of the orbit of the planet around the Sun. (In contrast to Tycho, who continued to be a neo-geocentrist, Kepler had from the beginning of his career been a Copernican.) The shape was, he found in 1605 (but published only in 1609), an ellipse, with the Sun at one focus. (Kepler was fortunate that he had based his analysis on Tycho's observations of Mars, since its orbit is eccentric enough for him to have made his great discovery. Had he begun, say, with Venus, with its nearly circular orbit, he would have failed.)

In the Ptolemaic scheme, Mercury and Venus had been placed between the Earth and the Sun. However, this was merely a convention. It would have been equally reasonable to place their orbits beyond the Sun. In the Copernican theory, the ambiguity of their position was removed – Mercury and Venus were clearly interior to the orbit of the Earth.

This implied something that had not been evident from the Ptolemaic point of view. Mercury and Venus could sometimes pass directly in front of the Sun, so as to appear as a moving black spot upon it. These passages, known as transits, occur at inferior conjunction – that is, when Mercury and Venus lie between the Earth and Sun, and turn their dark sides towards us. Usually, because Mercury's orbit is tilted by some 7 degrees to the plane of the Earth's orbit and Venus's by 3.5 degrees, they miss the Sun, passing above or below it. Only when an inferior conjunction takes place near one of the nodes – the points where the plane of the planet's orbit intersects the ecliptic – can a transit occur. There are two nodes (called the descending node and the ascending node, depending on whether the planet is above the ecliptic moving downwards or below the ecliptic moving upwards). The longitude of Mercury's descending node is the same as that of Earth on 7 May and that of the ascending node on 9 November, so transits of Mercury can only occur on or near these dates. On average, thirteen of them occur each century,

Statue of Tycho Brahe, with sextant, and Johannes Kepler, with a scrolled paper, by the Czech sculptor Josef Vajce, located on Keplerova Street in front of the Gymnasium in Prague. On this site was located the Kurz Summer Palace, where the two made observations.

which means they are infrequent but hardly rare. (For Venus, the corresponding dates are in June or December, but these are much rarer, and occur in pairs separated by more than a century. The last were in 2004 and 2012, and the next will not be until 2117 and 2125.)

Though a great deal of effort was later made to observe the rare transits of Venus, which became the centrepiece of a scheme to use observations from around the globe to work out the distance from the Earth to the Sun, transits of Mercury are interesting in their own right. Admittedly, because of Mercury's small size and greater distance from the Earth, they are hardly spectacular; Mercury (unlike Venus) is much too small to be detected with the naked eye (properly covered with eye protection, needless to say!) during a transit, and can only be seen using a telescope.

Even with a telescope, Mercury appears as a mere pinprick against the Sun – distinguishable from any sunspots that might be present by its inky black colour and perfect roundness of outline. Though Mercury is travelling in its orbit at a speed of about 170,505 km per hour (105,947 miles per hour), faster than any other planet, remember that the Sun is 1.4 million km (865,000 mi.) across; thus,

The Tychonic system of the World. In contrast to the Copernican system, Tycho's scheme kept the Earth at the centre, as in the Ptolemaic system, but while the Sun circled the Earth, the planets circled the Sun.

Pierre Gassendi's drawing of Mercury, at the transit of 7 November 1631. This event had been predicted by Kepler. Mercury was, however, so unexpectedly small that Gassendi at first thought he was observing a sunspot; it betrayed itself as the planet by its motion.

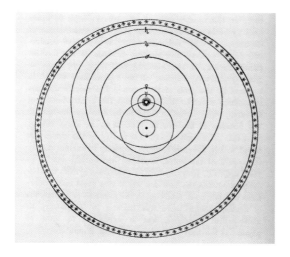

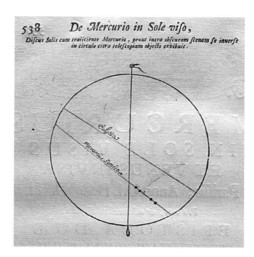

Mercury in transit across the Sun, 9 May 2016. Mercury is the small dark spot to the right of centre, which appears harder and sharper than the sunspots to the left of centre.

despite its high speed of motion, Mercury's progress is hardly perceptible from one minute to the next. It takes seven and a half hours to journey across the solar disc.

Kepler predicted, but did not live to see, a transit of Mercury on 7 November 1631.[3] Telescopes were then widely in use across Europe, and three astronomers observed the event, of whom the most notable was Pierre Gassendi (1592–1655), Roman Catholic canon of the parish church of Digne, in Provence. Gassendi, watching from Paris, was initially surprised by how tiny the planet was against the solar disc. At first he thought he was seeing an ordinary sunspot, such as had been discovered by Galileo and the Austrian Jesuit astronomer Christoph Scheiner (1573–1650) in the early days of the telescope. However, the spot's rapid motion soon convinced him that he was seeing the planet, and he wrote exultantly of his triumph to Wilhelm Schickard (1592–1635), professor of mathematics at the University of Tübingen: 'I have found him. I have seen him where no one has ever seen him before!'

Though the unexpectedly small size of Mercury in transit (13″ of arc (or 13 arcsecs), well below the threshold of naked-eye visibility) was one important result of the transit, another was the realization that such passages of the planet across the Sun provided an exacting test of the astronomical tables of the day. Instead of doing as astronomers had done from time immemorial – catching the planet as it twinkled low in the sky before sunrise and after sunset – they now had a sharp and precisely defined position against the Sun. From his observations, Gassendi calculated that the transit had actually occurred 4 hours, 49 minutes and 30 seconds ahead of the time Kepler had forecast. Given that nowadays such predictions are accurate to the fraction of a second, this may seem like missing the side of a barn; however, by the standards of the seventeenth century Kepler's prediction was extraordinarily good. His position for Mercury was in error by a mere 13′ of arc in longitude and only 1′5″ of arc in latitude. By comparison, the

Observatory of Hevelius at Danzig (now Gdansk), on the Baltic. The long telescope on a mast is an aerial telescope, which though impressive – it was 45 m (150 ft) in length – was clumsy to use. Most of the time Hevelius employed smaller telescopes, of which several can be seen located on the platform of the roof; with one of these he made out Mercury's phase in 1644. The observatory was largely consumed in the Great Danzig Fire of 1679.

then standard tables based on
Ptolemy and Copernicus were off
by an enormous 5 degrees – the
distance between the pointer
stars of the Big Dipper's bowl.

The next transit after
Gassendi's, on 9 November 1644,
went unobserved. The transit of 23
October 1651 (old style; 3 November,
new style) was observed by only
a single person, the English
astronomer Jeremy Shakerley
(1626–c. 1653), from Surat, India.
By contrast, the transit of 3 May 1661
– which occurred on the day Charles
Stuart was crowned King Charles 11
of England – was observed by a great
number of astronomers, including
the great Dutch astronomer and

Daniel Schultz, *Johannes Hevelius*, 1677, oil on canvas. Hevelius (Johann Hewelcke), was a brewer and City Councilman at Danzig (now Gdańsk), whose private observatory was the best equipped of his time.

mathematician Christiaan Huygens (1629–1695), who was then in
London, and the wealthy brewer and amateur astronomer Johannes
Hevelius (1611–1687), who observed it from his well-equipped private
observatory in Danzig (now Gdańsk, Poland).

The next transit of Mercury – and only the fourth observed since
the invention of the telescope – was that of 7 November 1677. Among
the observers was Edmond Halley (1656–1742), a 21-year-old student
at Oxford who had taken a leave of absence from his studies and was
then on the remote island of St Helena (where Napoleon was later
exiled), mapping the southern stars. His observations of the Mercury
transit showed him that the duration of the transit could be observed
exactly – he thought very exactly. He realized that, in principle, such
timings could be used to work out the distance from the Earth to the
Sun – though Mercury is too close to the Sun and too far from the

EDMVND. HALLEIVS LL.D.
GEOM. PROF. SAVIL. & R.S. SECRET.

Thomas Murray,
Edmond Halley, c. 1687.

Earth for observations to be taken with the necessary accuracy. Instead, he turned his attention to the much rarer but, at least for the purpose he had in mind, more favourable transits of Venus – an interesting story but one that lies outside our purview here.

Among Halley's most important other contributions was providing his assistance (and emptying his purse) in order to bring to press Isaac Newton's immortal *Principia,* setting out the theory of gravitation (first edition, 1687). Another was his brilliant adaptation of Newtonian methods in showing that the comet seen in 1531, 1607 and 1682 was one and the same, moving in a highly elongated orbit and returning to perihelion about once in every 76 years. (He predicted correctly that it would return again in 1758, but did not live to see it.) The Newtonian theory – in the hands of Newton's successors (many of them French) – led to increasingly accurate tables of the planets. Naturally it was expected that these improved tables would lead to near bull's-eye predictions of Mercury's transits as well.

But this was not the case. The prediction for the transit of Mercury of 1707 was off by a day, those based on Halley's own tables for the transit of 1753 were off by many hours, while even those of the celebrated French astronomer Joseph-Jérôme Lefrançois de Lalande (1732–1807) for the 1786 transit were wide of the mark by a miserable 53 minutes. Indeed, all the astronomers of Paris except one, Jean-Baptiste-Joseph Delambre (1749–1822), had already abandoned their telescopes and failed to see it at all. Clearly, something was amiss. In the nineteenth century the problem persisted, and frustrated the efforts of the greatest mathematicians of the heavens to solve it.

We will take up this thread again in Chapter Six, and also discuss the hypothesis generated from these small errors in Mercury's predicted and observed motions – the possibility that there might, in fact, exist another planet lying even closer to the Sun than Mercury. For now, we note only that the problem was eventually resolved. Nowadays there is no fear of astronomers missing a transit because of an error in the calculations. Indeed, the theory of Mercury's motion is now known with such great accuracy that it even includes a General Relativistic correction to the Newtonian calculations. Below is a list of twenty-first-century transits, of which, at the time of writing (January 2017), the first three have already taken place:[4]

7 May 2003
8 November 2006
9 May 2016
11 November 2019
13 November 2032
7 November 2039
7 May 2049
9 November 2052
10 May 2062
11 November 2065
11 November 2078
7 November 2085
8 May 2095
10 November 2098

Orbit

Mercury's year – the time it takes to travel once round the Sun – is only 87,969 Earth days. Its orbit has the highest eccentricity (0.206) of any planet – greater even than that of Mars. The mean orbital velocity of Mercury is 48 km/30 miles per second, increasing to 57 km/35 miles per second at perihelion and dropping off to 39 km/24 miles per second at aphelion.

At perihelion Mercury approaches to 46,001,200 km (28,583,820 mi.) from the Sun; but it recedes to 69,816,900 km (43,382,210 mi.) from the Sun at aphelion. The mean distance from the Sun is 57,909,050 km (35,983,015 mi.), or 0.387 of that of the Earth. (The Earth to Sun distance, 149,597,870 km or 92,955,807 mi., is referred to as 1.0 astronomical unit, or AU.)

Thus Mercury's distance from the Earth ranges between 82 million and 218 million km (51 million and 135 million mi.). Its distance from Venus ranges between 39 million and 178 million km (24 million and 111 million mi.).

As seen from the Earth, Mercury's apparent diameter is greatest at aphelic inferior conjunction, when its angular diameter attains 12″.9. However, it then lies almost on a straight line between the Earth and the Sun (and exactly on a straight line when a transit occurs), and its dark side is facing the Earth. At aphelic superior conjunction, it is on the other side of the Sun, and the angular diameter is only 4″.9.

The circumstances of Mercury's orbital motion and positions relative to the Earth and Sun will assume importance as we turn, in the next two chapters, to astronomers' attempts to observe it from Earth through telescopes. There we will consider the way Mercury appears from our particular vantage point in the Solar

Scale drawing showing the orbits of Mercury and the Earth round the Sun. The eccentricity of Mercury's orbit (0.206) is readily apparent even at this small scale.

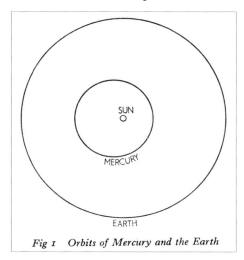

Fig 1 Orbits of Mercury and the Earth

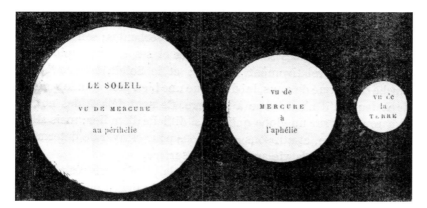

The apparent size of the Sun in the sky as seen from Mercury and the Earth: left, from Mercury at perihelion; centre, from Mercury at aphelion; right, from the Earth.

System. However, it is perhaps worth at least a moment to follow the nineteenth-century French astronomer-writer Camille Flammarion (1842–1925) and reflect on the way the Earth would appear to inhabitants (did such exist) on this little world, bound for ever to the solar neighbourhood:

> The inhabitants of Mercury see us shining in their sky at midnight as a splendid star of the first magnitude! Venus and the Earth are the two most brilliant objects in their starry nights. The Earth and Moon form to them a double star. If they have sufficiently powerful instruments, they may perhaps have already commenced to draw a geographical map of our planet – unless their religious and political principles affirm that Mercury is the only inhabited world, and prohibit the free examination of the heavens. The inhabitants of each planet have all, originally, believed themselves the centre of the universe, because they do not perceive their own motion, as the dwellers on earth do not perceive the motion of the Earth, and, like the Chinese, they declare they occupy 'the central empire', the rest of the universe being superfluous or given up to barbarians. Astronomy alone can upset these common illusions and lead the student to the Fountain of Truth.[5]

THREE

THROUGH THE TELESCOPE

As seen in a modern telescope, Mercury shows moon-like phases. A good 5–8-cm (2–3-in.) aperture telescope should suffice to show them. They were out of reach of Galileo Galilei's (1564–1642) small telescopes, of which the best was no more powerful than modern field glasses, but by 1639 an Italian Jesuit, Giovanni Zupus, claimed to make them out, and they were definitely seen five years later by the Danzig brewer and city councilman Johannes Hevelius.

The detection of phases for both the inferior planets Mercury and Venus – a term which does not suggest that they are somehow defective but that they lie 'below' or 'inferior' to the Earth and closer to the Sun, according to the way the ancients reckoned it – was important to seventeenth-century astronomers because of its bearing on the competing 'systems of the world'. According to the Earth-centred theory of Ptolemy, the inferior planets can show the new and crescent phases but never the gibbous or full. According to the theory put forward by Copernicus in 1543, the Earth was not at the centre but instead an ordinary planet circling the Sun. In this scheme, the inferior planets will show the whole gamut of phases from new to full. The detection of the phases of Mercury (and Venus) decisively ruled out the Ptolemaic theory.

It was not, however, quite an *experimentum crucis* establishing the Copernican theory. At the time a third scheme, the Tychonic system,

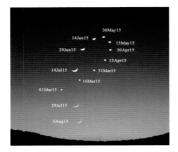 

The catch-22 of an inner planet's phases: the closer the planet comes to the Earth, the better the resolution, but at the same time the less of its Earth-facing disc is illuminated by the Sun. Here Venus is shown, but the principles for Mercury are the same.

was a serious rival. Put forward by the Danish astronomer Tycho Brahe, it placed the Sun and Moon in orbits round the Earth, while the rest of the planets travelled round the Sun, and was identical to the Copernican in the phases predicted for the inferior planets.

A plausible inference from the Copernican theory was that, if indeed the Earth were an ordinary planet (rather than the special and central seat of Creation), why might not some of the other planets be Earthlike, and have inhabitants of their own? The French writer Bernard de Fontenelle (1657–1757), writing in 1686, thought so, and in the case of Mercury, because of its proximity to the Sun, he speculated that the temperament of its inhabitants must be conditioned by their nearness to the Sun:

> They must be vivacious to the point of madness! I believe they have no memory . . . that they never think deeply on anything; that they act at random and by sudden movements, and that actually Mercury is the lunatic asylum of the Universe.[1]

Of course, at the time everything known about Mercury was entirely speculative; it was too small (not quite half again larger than the Moon), too remote and too poorly seen to show any detail. Apart from the phase, it was a cipher, and most observers quite understandably gave it only a cursory glance before turning to more rewarding fare. We know of no observations of Mercury by the great seventeenth-century astronomers Christiaan Huygens and Giovanni Domenico Cassini (1625–1712), who made so many wonderful discoveries about Mars, Jupiter and Saturn.

There was one circumstance in which Mercury – or at least its silhouette – could be examined for possible evidence of its physical condition. As noted in Chapter Two, Mercury and Venus occasionally transit the Sun. In Mercury's case this event occurs, on average, seven times each century (Venus's transits are much rarer), and it then appears as a small black dot moving across the brilliant backdrop of the solar photosphere.[2] The first transit of Mercury was observed at Paris in 1631 by Pierre Gassendi, canon of the parish church at Digne in Provence and a leading savant. We might say that it was the first transit of Mercury seen since the Creation, since the small disc of Mercury in transit is beyond the reach of the naked eye and can only be captured with the aid of the telescope. Other transits were observed in 1651, 1661, 1677 and 1697. At the transit of 5 May 1707, Abraham Sharp (1653–1742), assistant to John Flamsteed (1646–1719), His Majesty's 'observator' on Greenwich Hill, near London, noticed that the small planet appeared to be surrounded by a fuzzy bright ring, from which it seemed likely that the planet might be encompassed by 'a mist or dense atmosphere'.[3] This ring was seen by a number of later observers, including Johann Hieronymus Schröter (1745–1816) at his private observatory at Lilienthal, a small village near Bremen, during the transit of 1799. It was, he noted, a 'mere thought' in texture. Of course, as we now know, the effect has nothing at all to do with the planet itself: the bright ring is an optical illusion, produced by the eye-brain

Camille Flammarion's depiction of the bright aureole around Mercury observed during a transit, from *Les Terres du Ciel* (1884).

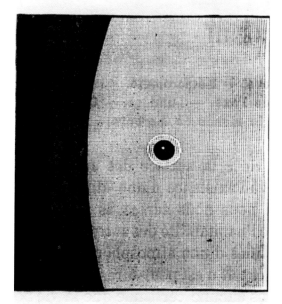

Fig. 168.
Auréole lumineuse observée autour de Mercure.

31

Transit of Mercury, 10 November 1973, based on observations by Richard Baum with an 11.4-cm (4 1/2-in.) refractor, magnifying 186x. The bright ring is an optical effect.

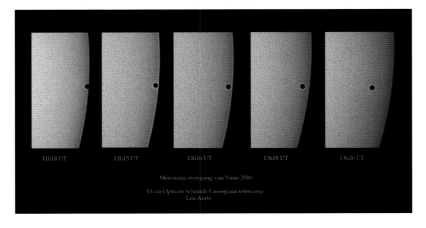

11h14 UT   13h15 UT   13h16 UT   13h18 UT   13h26 UT

Mercurius overgang van 9 mei 2016

13 cm Opticon Schmidt-Cassegrain telescoop.
Leo Aerts

Mercury transits the Sun on 9 May 2016. Note that even though these images were made by a highly skilful imager with a CCD camera, a bright, sharp ring appears around the black silhouetted disc of Mercury. Though once regarded as evidence of a Mercurian atmosphere, it is a purely optical effect.

system's inherent tendency to enhance contrasts at boundaries of shade. (It should be noted that images obtained with photographs or even Charge Couple Devices (CCDs) are not immune, since they are designed to produce images like those produced by the human eye.)

Schröter's Blunted Horn

Schröter, magistrate of Lilienthal, was one of the most enthusiastic amateur astronomers who ever lived. Unlike his great contemporary William Herschel (1738–1822), Schröter devoted himself to the systematic study of the Moon and the planets, and deserves to be called the 'father of lunar and planetary astronomy'.

Born in Erfurt (Thuringia, Germany) in 1745, he seemed destined to follow in the footsteps of his lawyer father, and in 1764 went to the University of Göttingen to study law. While progressing with his legal studies, he fell under the influence of the director of the observatory, Abraham Gotthelf Kästner (1719–1800), who awakened in him a lifelong interest in astronomy. In 1777 he moved to Hanover to take up a position as lawyer in the Royal Chamber of George III (this being the era of the 'personal union' of Great Britain and Hanover that lasted from 1714, the year George Louis of the House of Hanover ascended the throne of Great Britain as George I, to 1837, when Victoria ascended the throne of Great Britain). In Hanover, Schröter's musical inclination led to an acquaintance with the family of the regimental bandmaster Isaak Herschel. Schröter was befriended by Isaak's sons, Alexander and Dietrich Herschel, younger brothers of William Herschel. The latter was then pursuing a busy career as a professional musician in England while also developing into a keen amateur astronomer and telescope-builder. Contact with the Herschels reawakened Schröter's interest in astronomy, and through Dietrich he acquired his first telescope, a small achromatic refractor by the London optician Peter Dolland, which he was using by 1779.

The turning point of the lives of both William Herschel and Schröter occurred in 1781, when Herschel discovered the planet we now know as Uranus, and Schröter, in a spirit of emulation, decided henceforth to devote as much time as possible to astronomy. Perhaps it was something of a midlife crisis, for he was now 37. Instead of continuing with the demanding life of the Court, he left Hanover for the comparative idyll of Lilienthal, the 'Vale of Lilies' – a small town on the moor near Bremen. Here he successfully applied for a position as chief magistrate. Presumably his official duties were rather light, and he immediately set up the Dolland refractor in a small observatory he called 'Urania Temple' in the garden behind the court house. The small refractor failed to satisfy and, again making use of his contacts with the Herschel family, he acquired mirrors of 12 cm (4¾ in.) and 16.5 cm (6½ in.) fabricated by William Herschel himself. The latter, for which he paid a princely sum amounting to nearly half his annual salary at the time, was used to construct a 2-m/7-ft focal length telescope that was virtually identical to the one William Herschel had used to discover Uranus. When Schröter began to use it in 1786 it was the largest working telescope in Germany.

Schröter now began to publish, at short intervals and at his own expense, an impressive succession of astronomical works. His first involved sunspots. (It is not always remembered, incidentally, that he coined the term 'photosphere'.) He observed an outbreak of dark spots on Jupiter, and began the work for which he will always be known – on the Moon – in 1787. Originally intending to produce a map of the entire visible surface, he soon abandoned the project and satisfied himself with publishing a disconnected series of observations of specific features – the *Selenotopographische Fragmente* (Selenotopographic Fragments), of which the first volume appeared in 1791.[4] He also did valuable work on the planet Venus and, based on his sighting of a brilliant point standing apart from a blunted southern horn, he deduced the existence of towering mountains on that planet. Meanwhile, he was developing a large-telescope monomania. During 1792, he purchased another 2-m/7-ft reflector from Johann Gottlieb Friedrich Schrader, a physics professor at Kiel who in 1792–3 stayed with Schröter in Lilienthal and produced for him two more telescopes – one at 4 m (13 ft) with a 24-cm (9½-in.) diameter mirror and one at 7.6 m (25 ft). They may not have been of the best quality, and they failed to satisfy the ambitious astronomer, who in the meantime had taught his gardener, Harm Gefken, to grind mirrors. In the workshop next to the observatory Gefken fashioned a 47-cm (18½-in.) mirror for an 8.2-m (27-ft) reflector, which was set up in the open air and saw 'first light' by the end of 1793. It would remain the largest telescope in Germany for many years.

At that time, telescope mirrors were made of speculum metal, an alloy of tin and copper. Because of the copper, they tarnished easily, and in order to maintain the reflective surface Schrader had devised a method in which vaporized arsenic was added to the usual alloy. Sadly, Gefken was destined to become a martyr to his art: he died prematurely of chronic arsenic poisoning.

Inevitably, all of this telescope-making activity – and the cost of publishing his work – led Schröter into serious financial difficulties. Fortunately, George III came to the rescue, paying off all his debts provided only that he agree to leave his telescopes to the University of Göttingen when he no longer needed them himself.

As Henry King says in his *History of the Telescope*, 'Activities so unusual and so earnestly conducted could not fail to attract others.'[5] The others attracted included, in September 1800, several astronomers from across Europe who joined Schröter at his residence to found a 'Societas Liliatalica', devoted to scouring the stars of the zodiac for a planet presumed to lie in the gap between Mars and Jupiter. On the eve of the meeting, Mercury was appearing as an evening star, and Schröter turned the 48-cm (19-in.) reflector towards the shy rosy world. As he studied the wavering and usually tremulous image in his eyepiece, it flashed into sudden clarity from

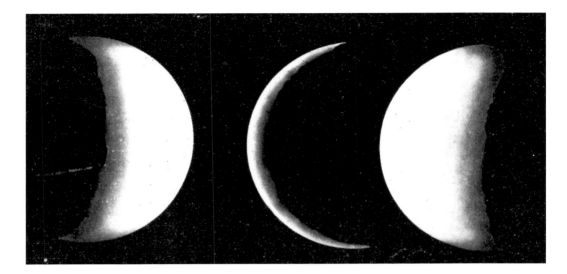

Schröter's drawings of Mercury. The one on the right clearly shows the blunted southern horn of Mercury, September 1800. Note that in this and in other telescopic observations, south is at the top.

time to time, and in these instants he suspected that the southern horn was blunted compared to the northern one. What he was seeing was no illusion: the impression was certain, and confirmed on subsequent nights. For that matter, his finding would be verified by many later observers of the planets, though his explanation – that the blunting of the southern horn of Mercury was due to a shadow cast by a mountain 18 km (11 mi.) high! – evoked scepticism. We will say more about the blunting of the southern horn later. What is significant at this point is that, equipped with a large telescope and impelled by his insatiable curiosity about other worlds, Schröter had succeeded in establishing a first toehold of factual information about the physical condition of the innermost planet.[6]

Since the blunted horn appeared more or less unchanged from one evening to the next, it suggested that the planet's rotation period might be almost in synch with the Earth's, that is, around 24 hours. Indeed, Schröter's assistant Karl Harding (1765–1834), who had come to Lilienthal as tutor to Schröter's children but soon became involved in the widespread observational programmes of the observatory, deduced from the observations a rotation period of 24 hours and 5 minutes, with an axial tilt of 20 degrees – both nearly identical values to the Earth's. A later Schröter assistant, Friedrich Bessel (1784–1846), reanalysed the observations and derived a rotation period even closer to that of the Earth – 24 hours, 00 minutes, 53 seconds. However, the axial tilt was revised to a most un-Earthlike 70 degrees.

It is tragic to record that even astronomers, gentle creatures though they are, do not escape the violence of natural and human disasters. The last decade of Schröter's life was disturbed by the Napoleonic Wars and the 'Vale of Lilies' was occupied by the French in 1806. Henceforth Schröter was cut off from his support from George III. Though he was given employment by the French, including the unenviable task of collecting taxes for them, he went unpaid. His income was now so limited that he was barely able to keep up the observatory, and his situation became even more desperate in 1810, when Napoleon decreed the assimilation of Lilienthal into the Département de le Bouche de Weser, centred in Bremen, and his position became redundant. Still, Schröter laboured on, and in 1811 he made extensive observations of the comet of 1811 (the one that would figure in Tolstoy's *War and Peace*).

The worst was still to come. In April 1813, as the French were reeling back after their ill-fated winter campaign in Russia, a skirmish took place near Lilienthal between a French detachment and a small band of Cossacks. A French officer wounded in the fray claimed his detachment had been fired on by the local peasantry, and without further ado, the commanding general, Vandamme, gave orders to set fire to Lilienthal in reprisal. A strong wind aided the conflagration, which destroyed the government buildings in which Schröter kept many of his books and manuscripts. Schröter fled with his family, 'in our night dresses', to his farm at Adolphsdorf, and though the observatory itself escaped the conflagration, several days later it was broken into and plundered by the French troops. They,

with a fury the most unprovoked and irrational destroyed or
carried off the most valuable clocks, telescopes, and other
astronomical instruments. My servants afterwards bought
a few of these articles [back] from the robbers; and I got back
a 3-foot achromatic telescope . . . The loss falls properly upon
the University of Göttingen, but I am willing to consider it
my own.[7]

With the expulsion of the French, Schröter was reinstated as
chief magistrate of Lilienthal, and attempted to rebuild the ravaged
village. But he was now a broken man, and it was too late for him
to rebuild the observatory. Sadly, with the help of his son, he began
transferring what survived of his instruments to the University of
Göttingen, and published his observations of the comet of 1811
and the 'fragments' on Mercury. He turned next to publishing his
observations of Mars, but his eyesight was failing, and the work was
published only in 1881. He died on the eve of his seventieth birthday,
on 29 August 1816. But his great work was done. He had established
on a firm basis the physical study of the Moon and planets, and in
addition had trained some of the greatest German astronomers of
the next generation, including Harding, who discovered the asteroid
Juno and became director at the observatory in Göttingen; Heinrich
Olbers, who discovered the asteroids Pallas and Vesta; and above
all Friedrich Wilhelm Bessel, who in 1810 took charge of the new
observatory at Königsberg (now Kaliningrad, Russia) and became
one of the greatest astronomers of the nineteenth century.

The durability of Schröter's constructions is nowhere better
attested than in the case of Mercury. For the next eighty years, few
would attempt to trespass the *ne plus ultra* laid down by the great
German pioneer. Now and again an observer turned a telescope
towards this unpromising object, and managed to make out a vague
marking of uncertain signification. But not until the 1860s did the
pace of Mercurian discovery begin to pick up.

For instance, in 1867, Charles Leeson Prince (1821–1899), a surgeon at Uckfield, Sussex, and a keen amateur astronomer and meteorologist, glimpsed an intriguing bright spot with faint lines diverging from it near the centre of the disc on three successive evenings. He derived a period of rotation of 24 hours, 5 minutes, 30 seconds – reassuringly close to Schröter's values. The noted British instrument-maker John Browning (1830–1925) in 1870 referenced observations of his countrymen Warren de la Rue (1815–1889) and William Huggins (1824–1910) of 'markings, like the lunar craters, of a dazzling whiteness' which appeared indistinctly 'as through a veil of mist'.[8] That same year an Irish amateur astronomer, John Birmingham (1816–1884), reported a large bright patch near the planet's limb. The year after that the German astronomer Hermann Vogel (1841–1907), then at the Bothkamp Observatory in Kiel, northern Germany, and best remembered today as a spectroscopist who made the first discovery of a spectroscopic binary star, detected bright spots on Mercury on two occasions with the observatory's 29-cm (11½-in.) refractor. With the same telescope, during a morning apparition in 1882, his colleague, Leo de Ball (1853–1916), caught glimpses of a dark curved streak as well as 'starry portions which seemed brilliant' near the planet's limb.[9]

The most important result of this era, but also one of the most unheralded, came from photometric work by another German astronomer, Karl Zöllner (1834–1882), which indicated that the planet had a very low reflectivity (albedo) and led him to conclude, 'Mercury is a body the surface condition of which must be nearly the same as that of our Moon, and which, like our Moon, probably does not hold an appreciable atmosphere.'[10] Previous astronomers, going back to Schröter, if not indeed to Flamsteed and Sharp, had assumed that Mercury was enveloped in an appreciable atmosphere. Zöllner's work thus marked a new departure. Unfortunately, he published in an obscure journal (*Poggendorff's Jubelband*), where his work attracted little notice. Later observers would pay dearly for their failure to take heed of his findings.

At this point, we remark what can be deduced from a comparison of Mercury with another planet, on the rare occasions when they are situated in the same field together. An opportunity of this kind came on 28 September 1878, when Mercury lay in the same field as Venus, and the two planets were studied by the Scottish foundry man turned amateur astronomer James Nasmyth (1808–1890). He noted that Mercury was as 'dull as lead or zinc' compared to the 'clean silver' of its dazzling, cloud-shrouded companion.[11] I observed a similar encounter of these same planets, with historian of astronomy Owen Gingerich, from the roof of the Science Building at Harvard University on 28 June 2005. Nasmyth's comparison seemed apt, except perhaps for requiring a little copper to add to the lead or zinc. The phases were gibbous and very similar at the time, and Mercury was extremely dull compared to its dazzling companion.

FOUR

ROTATION

Early in the 1880s, two astronomers, one amateur, the other professional, made a renewed attempt on the innermost planet. One engaged in a brief skirmish, the other in a prolonged siege. The amateur was William Frederick Denning (1848–1931).[1] In 1871 Denning acquired a 25-cm (10-in.) altazimuthally mounted With-Browning reflector, which was to become the main instrument in his researches on the planets. At first his main effort was directed at Jupiter, whose Great Red Spot rose to prominence in 1878. At the same time he became keenly interested in meteors, and his expertise was acknowledged in H. G. Wells's 1898 novel *The War of the Worlds*. When the first of the Martian cylinders drops to the Earth, Wells writes,

> Denning, our greatest authority on meteorites [*sic*], stated that the height of its first appearance was about ninety or one hundred miles. It seemed to him that it fell about one hundred miles east of him.[2]

Denning was also a keen observer of the other planets, and during its morning elongation in November 1882 Denning turned the 25-cm (10-in.) reflector on Mercury as it raced ahead of the Sun. Denning once recorded his delight in catching sight of the elusive planet – though at one of its evening rather than morning elongations – in a smattering of verse published in his classic book *Telescopic Work for Starlight Evenings*:

Come, let us view the glowing west,
    Not far from the fallen Sun;
For Mercury is sparkling there,
    And his race will soon be run.
With aspect pale, and wav'ring beam,
    He is quick to steal away,
And veils his face in curling mists –
    Let us watch him while we may.[3]

At the November 1882 morning elongation, Denning was able to follow Mercury into the brightening twilight by carefully turning the manual slow-motion controls of the telescope's mounting and, as the planet ascended above the thickest and most disturbed layers of the Earth's atmosphere, was rewarded with moments of 'extreme sharpness' in which, as he recorded in his observing log for 8 November, he distinguished 'a very brilliant small spot, with luminous radiations extending over the whole area' near the centre of the half-illuminated disc. It was a tantalizing case of what Percival Lowell (1855–1916) would refer to as a 'revelation peep'.[4] (In the spacecraft era, it would emerge that in this 'revelation peep', Denning had, very likely, caught a glimpse of the extensive ray system associated with the crater now known as Degas.[5])

William Frederick Denning, whose pioneering observations of the planets, including Mercury, were made with a 25-cm (10-in.) With-Browning reflector on a simple altazimuth mount.

Denning's success would lead him to state on a subsequent occasion that the perception of the planet as a difficult and unrewarding object was 'partly induced by a misconception'.[6] Observers' defeat was owing to their failing to observe it at any time other than when it was at very low altitude, and its image violently disrupted by air currents. He had realized this in November 1882, when he had followed it ahead of the Sun into higher elevations,

Mercury in 1882.
Observations by Denning
with a 25-cm (10-in.)
With-Browning reflector.

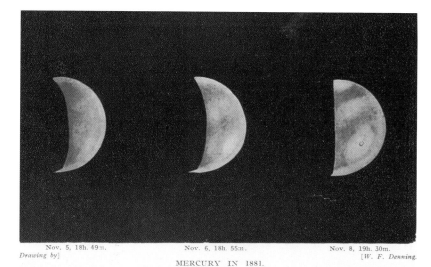

Nov. 5, 18h. 49m.
Drawing by]

Nov. 6, 18h. 55m.

Nov. 8, 19h. 30m.
[W. F. Denning.

MERCURY IN 1881.

For many reasons Mercury is a difficult object for telescopic study, but, under favourable conditions, markings are definitely observable on his surface. They generally take the form of dusky bands or patches, but can seldom be seen and identified often enough to give a reliable value of the planet's rotation.

and so enjoyed more favourable conditions. Indeed, he claimed the markings on the planet were 'so pronounced that they suggest an analogy with those of Mars',[7] while E. M. Antoniadi credited him with being 'the first to show genuine patches' similar to those seen by later observers of the planet.[8]

We now know that Denning had indeed captured genuine features of the planet's surface – including, as noted above, the prominent ray system associated with a crater. However, he subscribed to the belief that Mercury must be surrounded by a dense atmosphere, and concluded that 'faint, irregularly shaped, dusky spots and white areas' were probably atmospheric features, having 'no durable character'.[9]

After achieving this success, Denning wrote to another astronomer who was also studying Mercury at the time: Giovanni Virginio Schiaparelli (1835–1910), a professional astronomer and director, since 1862, of the Royal Observatory of Brera in the Brera Palace in Milan. In fact, Schiaparelli had already been chasing the

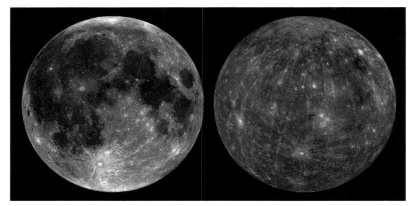

The Moon and Mercury compared in images taken by the Lunar Reconnaissance Orbiter and MESSENGER 10 space probes, respectively.

innermost planet since the previous January, observing it, whenever practicable, from his dome atop the palace.

Mars seen through a small telescope from the Earth. Drawing by Richard Baum, 1 January 1993, with an 11.4-cm (4.5-in.) refractor, 186x. At the time Mars was 93 million km (58 million mi.) from the Earth.

In his letter to Schiaparelli, Denning made the interesting observation that the general appearance of the Mercurian disc was similar not to the Moon, as might have been expected, but to Mars. It was a thought-provoking comparison, given that the Red Planet was just then being catapulted – largely as a result of Schiaparelli's discovery in 1877 of the *canali*, or 'canals', crisscrossing the surface of that planet – into the foreground of the public's attention as the world beyond Earth most likely to have inhabitants. Perhaps this remark was partly inspired by the rosy colour of the disc, occasioned by the planet's low altitude in the pre-dawn sky, but it may also testify to the fact that observers' perceptions of Mercury were rather unstable at the time, teetering precariously between images of the Moon and the Earth, with Mars perhaps offering the most viable compromise between the two.

MARS - 1993

Jan. 1ˢᵗ 2030 -2102 UT.   115mm OG x186
ω = 300°;   So. Ⅱ.   Trans. 4/5.   FROST.

Cloud over Isidis Regio. Morning limb bright.

After November 1882, Denning retired from the field. (He does not appear to have observed Mercury again until 1906.) In this respect he was like all previous students of the innermost planet who had watched it through one or two elongations before going on to more rewarding game. Schiaparelli, on the other hand, was embarking on nothing less than a prolonged siege. In fact, he would devote eight years to this difficult endeavour.

An Accidental Planetary Astronomer

Though Giovanni Virginio Schiaparelli ranks among the most skilful visual planetary observers of the nineteenth century, he came to this branch of the science relatively late, and after already amassing an impressive array of accomplishments in other directions.

Born in Savigliano, a small town in the Piedmont region of northern Italy, on 14 March 1835, Schiaparelli's father was, like his forefathers for centuries, a kilnsman, a maker of bricks and tiles. Though the family was not well off, they were strongly committed to the education of their children, many of whom possessed great talent and went on to careers of distinction.[10]

Thus, at age four, Giovanni was taken outside by his father for a stimulating look at the stars. He would long remember the excitement about the heavens, which planted the seed that would grow into his life work:

> Thus, as an infant, I came to know the Pleiades, the Little Wagon, the Great Wagon . . . Also I saw the trail of a falling star; and another; and another. When I asked what they were, my father answered that this was something the Creator alone knew. Thus there arose a secret if confused feeling of immense and awesome things. Already then, as later, my imagination was strongly stirred by thoughts of the vastness of space and time.[11]

His interest was further stirred by the total eclipse of the Sun
on 8 July 1842, which he viewed through the window of the family's
casa. He also benefited from the instruction of a learned priest, Paolo
Dovo, who lent him books and gave him his first views through a
telescope of the phases of Venus, the moons of Jupiter and the rings
of Saturn, from the campanile of the church of Santa Maria della
Pieve. The youngster proved to be an excellent scholar, and after
learning everything he could from the local schools in Savigliano,
he was sent to the University of Turin, where he graduated with
a degree in architecture and hydraulic engineering.[12] Already,
however, he was utterly in thrall of astronomy. After a brief stint
as a schoolteacher in Turin, a position for which he seems to
have been rather ill suited, he was granted a stipend from the
Piedmontese government allowing him to receive training in
astronomy at two of the leading observatories of the day, those
at Berlin and Pulkovo.

In 1859, he was recalled to Italy. At the time the struggle
for independence from Austria – a struggle in which the
Piedmontese were taking a leading role – was still in the
balance.[13] Seeing it in part his patriotic duty, in June 1860
Schiaparelli accepted the position of *secondo astronomo*, under
the aged director Francesco Carlini, at the Brera Observatory
in Milan. He arrived at Brera within a month of the Expedition
of the Thousand under Garibaldi, which would prove to be
the turning point in the war for Italian independence, and
after Vittorio Emanuele II of the House of Savoy assumed
the throne of the United Kingdom of Italy on 14 March 1861,
Schiaparelli discovered politically well-connected allies in
Savoy who were willing to pull strings for him. Their string
pulling was successful, and so at the age of only 26, Schiaparelli
was appointed director of the Brera Observatory – now the
Royal Observatory. Despite an unsuccessful attempt to lure
him away to the Arcetri Observatory in Florence, to fill the

In this impressive bronze
sculpture by the painter
and sculptor Annibale
Galateri from 1925, which
graces Schiaparelli's home
town of Savigliano, the
great astronomer continues
to ponder the mysteries of
the heavens.

place left on the death of the founder, G. B. Donati, Schiaparelli remained at Brera for the rest of his life.

The observatory had been founded in the old Brera Palace by the Slovenian Jesuit Roger Boscovich (1711–1787) a century before, and possessed what was then an interesting but rather useless collection of ancient instruments. Though Vittorio Emanuele II was a sot and, according to the English diplomat Lord Clarendon, an 'imbecile', whose main priorities were paying off mistresses and keeping up hunting lodges, most of which he never visited, the minister of education, Quintino Sella, was an intellectual and a supporter of science. Like Schiaparelli himself, Sella had been educated in engineering at the University of Turin, and he managed to wangle funds for the Brera to acquire a first-rate telescope. It was, however, a long time in coming. Only in 1874 was the instrument Schiaparelli would make famous, a 22-cm (9-in.) Merz refractor,

The dome of the 22-cm (9-in.) Merz refractor on the roof of the Palazzo Brera, as it looks today. The bell tower of the church of San Marco stands in the background.

finally installed in a dome on the rooftop of the palazzo. For two years Schiaparelli put it to work measuring close double stars – work which he would continue, systematically, for the next 25 years.

This refractor – removed long after Schiaparelli's time and exhibited as a museum showplace, during which time all the brass and other moveable parts were pilfered by thieves – has recently been beautifully restored, and once more stands under its copper cupola. It is a sleek, elegant instrument with a polished mahogany tube and a new mounting, a motorized drive mechanism having been substituted for the original, which was regulated by a weight-drop clock mechanism. The objective, crafted by Merz, is of course the same. The crown element of the achromat is lightly greenish stained, owing to iron impurities in the glass, as was already the case in Schiaparelli's time. The chromatic aberration is slightly overcorrected so that the objective is achromatic in the red-green region of the spectrum (perfect for Mars!) but produces an excess of blue. Ever the innovator, Schiaparelli corrected this effect by employing yellow or orange filters. Among other techniques he pioneered were illuminating the field of the eyepiece to improve the contrast of planetary surface details and – most relevant here – observing the planets in broad daylight.

This, then, is the telescope with which Schiaparelli made, as Percival Lowell expressed it, 'voyage after voyage, much as Columbus did on Earth'.[14] They were voyages of discovery that opened up new vistas on the Martian world – and also on the Mercurian one.

Until 1877 Schiaparelli seems to have had no plans to use the telescope for such voyages of discovery. That year he turned 42, and was best known for an investigation he had carried out in his early thirties. In a brilliant bit of sleuthing, he had demonstrated that the swarms of particles that produce 'meteors' when they burn up in the Earth's atmosphere follow the orbits of certain periodic – that is, regularly returning – comets. Thus he

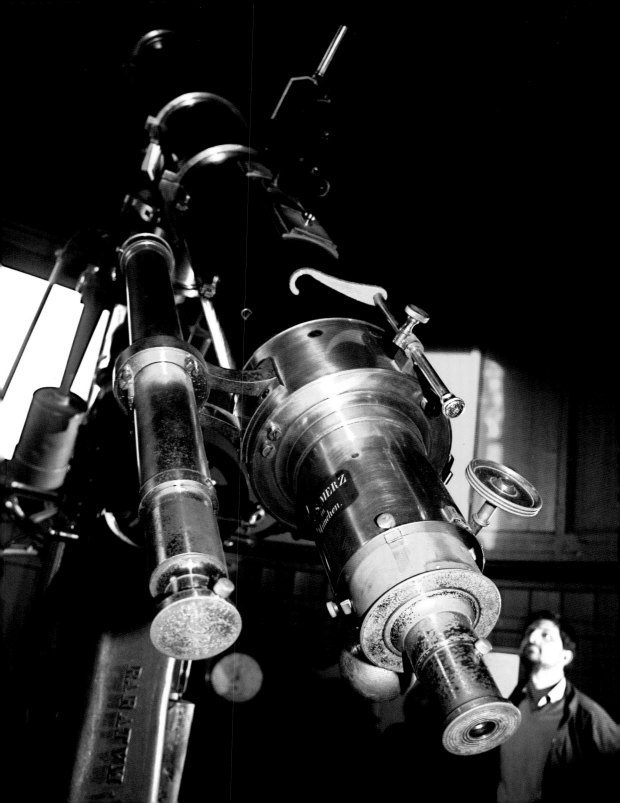

had answered the query about falling stars he had put to his father as a four-year-old.

On 23 August 1877, Mars was a brilliant object, only two weeks from making an unusually close swing by the Earth (56 million km, or 35 million mi.). That night there was an eclipse of the Moon, which Schiaparelli observed. Mars lay close by, and to pass time during the relatively uninteresting partial phases of the eclipse, Schiaparelli decided to test the optical properties of his telescope by turning it on the planet. He had not been prepared for the subtle beauty of the surface markings, and on subsequently researching previous attempts to map their forms and positions he realized that none was entirely satisfactory. Then and there he decided to produce his own map. On that map, which appeared a year later, he introduced a new scheme of Martian nomenclature based on the geography of the ancient classical world (it is, by the way, the basis of that still in use today), and also introduced to the world the strange linear markings he called *canali* (or, as they soon became known in the English-speaking world, 'canals'). The 'canals', which some (though not necessarily Schiaparelli himself) thought artificial, soon became the centre of a worldwide controversy sometimes referred to as the 'Mars furore', that would rage for the next thirty years and continue to reverberate for decades beyond that. Schiaparelli himself continued to observe the planet. He produced a new map at the next Martian opposition, in November 1879, but before the one after that, in January 1882, he had taken on board another subject on his programme of planetary reconnaissance: Mercury.

Mercury at Milan

Schiaparelli's first attempt at Mercury occurred in June 1881, at the height of Milan's sweltering summer heat. On opening the shutter of the dome on the roof of the Palazzo Brera, he used the telescope's setting circles to point to an object utterly submerged

The refurbished Merz refractor, as it looks today. After long languishing as a museum showpiece, it has returned to its dome at Brera and, beautifully restored (in part with a grant from the Schiaparelli family), is once more in working order.

into the glossy blue Milanese sky and invisible to the naked eye. Having already arrived at the conclusion that Denning would reach more than a year later – that Mercury was a hopeless object if viewed only during the usual twilight intervals, when it appeared to the naked eye as Stilbon, the 'scintillating one', and in the telescope as a boiling, coruscating, confused mass of light and colour vexed by air currents – Schiaparelli hit upon a novel plan: what if it were possible to flush the planet from its hiding place in the daylight sky and study it while much higher above the horizon? This first test showed him that it was possible. The planet appeared pale but distinct against the blue-sky background. With a magnifying power of 200x – enough to give an image about two-thirds the size of the Moon seen with the naked eye – he could make out various spots or shadings. They were delicate, however, and completely erased whenever a slight haze or layer of cirrus clouds interposed themselves. Schiaparelli's experience with Mars had shown him that the Milanese summer was not the time to push forward with this research, however. As he wrote:

> In our latitude, useful observations of Mercury at night are impossible, while even at twilight conditions are generally unfavourable, since the planet's low altitude prevents the use of magnifications necessary for the purpose, which ought to be at least 200. In order to obtain a good series of observations, I therefore decided to attempt to do so with the Sun still above the horizon, which, when the air is pure and calm, is feasible at any hour of the day in the winter, and in the early morning hours in autumn and spring. The observations are more difficult in the summer, because of the vapours which arise from our damp plains, and more so in consequence of the almost perpetual tremors in the atmosphere caused by the strong heating of the ground and of the structures of the vast city surrounding the observatory on every side.[15]

Having satisfied himself 'that it would be possible not only to see the markings on Mercury in full daylight, but also to obtain a series of sufficiently connected and continuous observations of these spots', Schiaparelli put the matter out of his mind until January 1882, when conditions would be more favourable for his project.

That January – as during most of his working career as an astronomer – Schiaparelli no doubt did as he always did, and made the five minutes' walk from his house, at 7 Via Fatebenefratelli, to the Brera, continued through the colonnaded courtyard, and went up the (right) staircase to his office. (The left staircase led to the celebrated Brera Pinacoteca, whose masterpieces include works by Raphael and Caravaggio.) He was a creature of habit, and only with great reluctance could he ever be dislodged from strict routine. He had been well travelled as a student, but after his appointment to Brera he made only one trip outside Italy – his honeymoon to Vienna; reportedly it did not go well.[16] In his logbook, recording 25 years of double-star observing, he regretted having to leave the observatory fourteen different times. Several trips were to the Quirinal Palace in Rome, where he was summoned to present his discoveries to King Umberto I and Queen Margherita. One was to take his oath on being appointed by decree a senator in the Kingdom of Italy. Though he showed up to prove that he was a good patriot, he did not believe he was suited to being anything more than an honorary senator: 'Senators make laws for men,' he said, 'but I only know a few of the laws of the skies.'[17] The great astronomer maintained a wide correspondence, but preferred to work alone. Thus he never took students, and whenever he was introduced to a stranger, he 'bristled like a hedgehog'.[18] During his entire career as an astronomer, he typically spent at least ten hours per day in his office, divided into three intervals. During the intervals he presumably went home to spend time with his wife and large family. When he planned to observe at night, he skipped lunch, but slept a little before going up to the telescope, for he insisted that in order to observe well, the brain and eyes must be rested. He never smoked, drank little coffee or wine, and in general abstained from anything that might be upsetting to the nervous system.

As he began his study of the little planet he was soon to be affectionately referring to in his observing logbook as his *amici*, his friend, Schiaparelli above all hoped to determine an accurate value of the planet's rotation. (Schröter's Earthlike period of around 24 hours had stood for eighty years without a serious challenge.)

Though he took great care in sketching the surface markings day after day, he tried – as we will see, not entirely successfully – to refrain from prematurely guessing what the observations might be telling him. He clearly meant the drawings and notes in his logbook to serve as raw data. Accordingly, his notes seem aseptically clinical: they refer blandly, not to say tediously, to spots shown in the drawings, which he indicated not with romantic names from antique geography (as with the surface features he had mapped on Mars) but perfunctorily with lower-case Latin letters.

Schiaparelli was in some ways an unlikely visual observer of the planets. He had only one good eye – his left; the other was congenitally defective. He also suffered from red-green colour-blindness. At least with a planet of mere differences in half tones like Mercury, this may have been an advantage, since it seems to have made him particularly sensitive to subtle boundaries of light and shade. (My radiologist colleagues tell me that for the same reason they routinely employ grey scale for their readings.) Though Schiaparelli's notes about planetary colours thus need to be taken with a pinch of salt, he described the shadings as reddish-brown, and the background disc as bright rose to coppery.[19] Though observing in broad daylight conferred advantages in the steadiness of the images and the duration of time in which they could be studied, there were trade-offs as well: contrast suffered compared with the twilight periods, and under these conditions the markings always appeared washed out, as 'extremely delicate dark streaks, which under normal observing conditions could be recognized only with much effort and great concentration'. And 'all these streaks are clear brown in colour against a rosy background, always smoky, not easily seen against the background, and difficult to distinguish.' Nevertheless, they were there; they were not illusions. Only when tremors of the atmosphere or poor sky transparency obliterated them were they not present at all. The problem was, Schiaparelli found, not so much in seeing them as in sketching accurately what he had seen. As he wrote: 'It is most difficult to give a satisfactory graphic representation of such vague and diffused forms or bands especially from the want of fixity of the edges which always leaves room for a certain choice.'[20]

The Testimony of Observing Logs

Schiaparelli's first sketch was entered into his observing log on 27 January 1882. At the time, the planet was like a small gibbous moon approaching a 6 February elongation east of the Sun (evening apparition).[21] He continued this first series of observations until 10 February, making no fewer than thirty sketches in a period of just over two weeks. (His sketches of Mercury appear alongside sketches of Mars, which he was observing intensively around the time of its opposition.) In his first attempts at the elusive inner planet, the markings are very vague and indistinct. Finally, on 4–6 February – with the image in the telescope resembling a small half-Moon – Schiaparelli's attention was seized by a feature more arresting than any of the others.

The distinctive feature consisted of a system of spots (labelled w a b k i), which looked for all the world like the Arabic numeral 5. Schiaparelli continued to remark on its presence over the next several days. The feature lay in the illuminated region west of longitude 90 degrees (according to modern reckoning). It was a definite and unmistakable form – a Jamesian figure in the carpet, more regular and clear cut than anything seen so far, and furnishing him with a reference point. As we shall see, it would also ensnare him in a trap. During all of his later observations of the planet during its elongations east of the Sun, he would be haunted by this hallmark figure.

It was Schiaparelli's intention, right from the outset, to keep Mercury under as continuous scrutiny as possible. If it was, as a seventeenth-century writer described, like a debtor hiding from creditors, he was determined not to let it get away. After passing through inferior conjunction on 22 February 1882 (when it was invisible between the Earth and Sun), Schiaparelli caught up with it again on 10 March, as it re-emerged from the solar glare. It was now headed towards a greatest elongation west of the Sun, and thus appearing as a morning star. He continued to observe it until 27 April, making sixteen sketches during this western apparition. The most prominent feature seen during this series of observations was a marked dusky spot, which he labelled 'q'.

As the planet came to its next eastern elongation (1 June 1882), Schiaparelli had kept it under scrutiny through an entire synodic

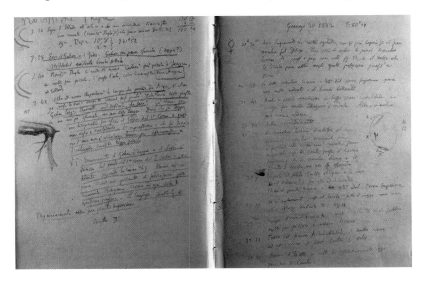

Pages from Schiaparelli's observing logbook. His drawing of Mercury on 30 January 1882 is on the right-hand page, while the left-hand page contains a sketch of the Sinus Sabaeus and associated 'canals' on Mars.

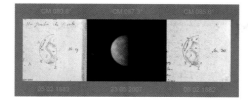 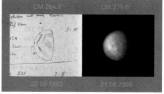

The figure-of-5 makes an appearance. These sketches from Schiaparelli's logbooks, made on 4 and 6 February 1882, record the feature that was to mesmerize and mislead him throughout his long campaign of observations. The Central Meridians for the dates in question, according to modern reckoning, are given. A CCD image by John Boudreau under very similar conditions is shown for comparison.

Mercurian mind games. Compare the drawings of February 1882 with his drawing one synodic period later, on 22 May 1882. A Boudreau image is shown for comparison. Schiaparelli believed he was seeing the same markings – the figure-of-5 – when in fact he was looking at opposite hemispheres of the planet.

period (116 days). In the interval between 5 May and 5 June, he completed fourteen sketches in all. The lesser number of drawings reflects the arrival of the hot Milanese summer, and a consequent deterioration of the astronomical 'seeing'. Indeed, Schiaparelli noted that sometimes the cupola on the roof of the Brera Palace became 'infernally hot' and that the column of air above the telescope was in a continual boil.[22] Still, he saw – or seemed to see –the figure-of-5 again. What had begun as a singular detection was quickly settling into a fixed stereotype: the figure-of-5 was, he would assert, 'characteristic at the greatest elongations east of the Sun whenever the disc of Mercury is seen half-illuminated'.[23]

Rather naturally, he concluded that, since he was seeing the figure-of-5 marking under similar conditions of phase, the apparition was one and the same with that seen in February. Perhaps he ought to have been more cautious given the difficulty of the observations. However, his conclusion was not unreasonable. As we now know, however, in fact he had taken a wrong turn: he was actually now examining features on the opposite side of the planet, which had completed two rotations in the interval between observing windows. His assumption that they were the same was an example of a 'bottom-up' influence on perception: figure-completion.[24]

In addition, there was also another likely influence on Schiaparelli's perception – a top-down influence (sometimes referred to as 'innocent expectation' or the 'beholder's share'). Though he gives no hint in his observing logbooks, which are mere bland registers of fact, he was no doubt aware of interesting and

topical work on the evolution of the Earth-Moon
system, published only a few years earlier by the
British astronomer George Darwin (1845–1912).
In 1877 Darwin (son of the famous naturalist) had
shown, by means of an exhaustive mathematical
study, how tides raised by the Earth on the primordial
Moon had gradually slowed the latter's spin until
it had come to turn on its axis in the same period
of time that it takes to complete one orbit around
the Earth. Darwin himself coined the terms 'tidal
friction' and 'captured rotation' to describe the
mechanism and its consequence. It was natural to
suppose that the same mechanism had been at work
on Mercury, where the immense tides raised by the
Sun should have braked its rotation with sinister efficiency until
it became equal to its year of 88 (terrestrial) days.

Sir George Howard Darwin.

   With these 'bottom-up' and 'top-down' influences working in
combination on his perceptions, Schiaparelli observed through the
entire summer of 1882. His campaign on Mercury even included a
heroic – or reckless – attempt in August to chase the planet to within
only a few days of superior conjunction. Though he found he could
sketch the planet under these conditions even without taking special
precautions for eliminating the solar rays from the telescope, he
would later admit that these 'dangerous' observations had probably
seriously damaged his eyesight.[25] His logbook entry on 11 August
describes Mercury only 3°2′ from the Sun:

> Under such circumstances the disc of the planet appears almost
> perfectly round, with the light only a little less than uniform; but
> despite the fact that the apparent diameter was reduced to 4″ or
> 5″, the positions of the observable markings could be judged
> with greater certainty than at other times.[26]

The next greatest elongation east of the Sun (Mercury as an evening star came on 28 September 1882, and would provide him with his *experimentum crucis*: if the figure-of-5 again appeared, he would have a lock on the synchronous rotation. The result: like 'Old Faithful', there again was the figure-of-5.

Schiaparelli had by now reached a quite unshakeable conclusion. Thus on 20 October 1882 he confided to his closest astronomical friend, François Terby (1846–1911) of the University of Louvain, 'I believe that my researches . . . are advanced enough to give you a first idea of my findings. If I should happen to die before I publish them, I pray you will do so, so that this beautiful result will not be lost to science.'[27] And what was the beautiful result? In the tradition of earlier astronomers, Schiaparelli communicated it to Terby in a Latin verse (translated here):

Cyllenius, turning on its axis after the manner of Cynthia,
Eternal night sustains and also day:
The one face burned by everlasting heat,
The other, hidden, never seeing sun.
Be not Ceylon by you admired more,
Which Titan, fiery potent, burns with his rays,
Nor the Riphaean mountains, paralysed with cold,
Nor Thule, sunk in the night of the Bears' heaven.
Luna, to be sure, in her great changes is scorched and freezes,
And reckons 'days' what you measure as 'months'.
But the wretched star that spins in the first circle
By greater flame, by greater ice, is touched.[28]

Mercury, shown in Schiaparelli sketches of 7 and 8 August 1882, as it was approaching superior conjunction and thus drawing to within only a few degrees of the Sun in the sky. Despite the brightness of the daytime sky background, Schiaparelli found the markings distinct: the prominent marking labeled 'q' is the same that he thought he had seen when Mercury was near its greatest elongation west of the Sun the previous April, while what he thought to be the figure-of-5 is shown toward the left side of the disc. A Boudreau image is shown for comparison.

Mercury, almost full, captured by Schiaparelli near superior conjunction, when it was only 3° 2′ west of the Sun. Compare the sketches. Schiaparelli again records the prominent q marking, while the figure-of-5 shown in the previous set has presumably librated off to the left of the image. A Boudreau image is shown for comparison.

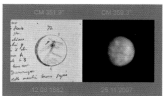

More prosaically stated, Schiaparelli believed he had obtained a result that was satisfyingly consistent with Darwin's theory of tidal friction: Mercury's period of revolution and rotation were synchronous – the periods were the same, 88 terrestrial days or, to be precise, 87.96926 days.

Schiaparelli kept up his observations with the same enthusiasm – or diligence – for another year. By September 1883, he had made in all 134 drawings of the planet, providing coverage through a full seven synodic periods. In November, he wrote again to Terby to account for a long and uncharacteristic lapse in communication, which he attributed to 'the unhappy combination of heavy work with a decrease of health and vigour, obliging me to lay aside my correspondence as well as several other works to which I attach interest. Must I confess to you that for several months I have done little more with Mercury?'[29] There was no reason to doubt him; he was frequently ill, probably mostly owing to overwork. However, he was also struggling with Mercury, and withheld publication. The fact is there was tarnish on his triumph, as he was finding that at times when the characteristic features of the planet's evening and morning faces – the figure-of-5 and the 'q' respectively – ought to have been visible, they were either difficult to identify or altogether absent. Some variability was only to be expected given the changes in the transparency of the atmosphere of Milan, while others could be accounted for by the effects of libration: as the planet's velocity varied along its highly eccentric orbit, the constant rotation got out of step with its orbital motion, as had long been noted in the case of the Moon. The planet would thus appear to rock slightly back and forth, and Schiaparelli calculated that the extent of this excursion was considerable, amounting to some 23 degrees in longitude and 7 degrees in latitude. Thus a marked variation in the visibility of the surface features – and, particularly, the reason that the markings seemed to fade out periodically towards the eastern and western limbs – could be attributed to this cause. The librations also had another – intriguing – consequence. Though part of the surface would be scorched by perpetual day and part frozen by eternal night, there would also be an intermediate zone of alternating sunrises and sunsets. Comprising about a quarter of the planet's surface, this 'twilight zone' might experience variable seasons and even – so he would later tell the king and queen of Italy – support life.

The librations only partially resolved his difficulties, however, for there were times when the atmosphere over Milan was perfectly transparent and steady, and yet the surface features were unaccountably vague; at other times parts of well-known configurations like the figure-of-5 were missing. There were even times when everything was confused, as if covered by a dense veil. There seemed to be only one explanation, and Schiaparelli may have come to it rather reluctantly. According to the kinetic theory of gases just being formulated at the time, the atmosphere around a small planet like Mercury, so close to the Sun, would have been expected long since to have boiled away into space. The same conclusion seemed to follow from Zöllner's photometric measures, which showed that Mercury's albedo was as low as the airless and moistureless Moon's. But Schiaparelli was a classical astronomer, and not very interested in physics; he may not have kept up on the kinetic theory of gases, nor does he seem to have read – or if he read to have remembered – Zöllner. Convinced by the end of 1882 – and still convinced by the end of 1883 – that Mercury's rotation was synchronous and that the planet was tidally locked with the Sun, there seemed to be only one way to 'save the phenomena'. In addition to the variability of the markings, he called attention to various 'white spots' seen from time to time. 'These occur,' he wrote,

> for the longest periods near the limb of the planet, where they sometimes become very splendid; but they are not rare even on the inner parts of the disc, only they are then less bright and harder to recognize . . . Above all they occur in the region . . .

near the north pole. It may not be too rash to suppose that
these white veils, and also the variable intensity of the dark
spots and their tendency to disappear near the limb, are
due to more or less opaque condensations, produced in the
atmosphere of Mercury, which from afar presents analogous
aspect to what the Earth would show from a similar distance.[30]

We shall see later what to make of these 'white spots' and 'white
veils'.

Though Schiaparelli had reached these conclusions by the
end of 1883, he did not publish right away. He was a cautious
individual, and decided to wait until he could confirm his work
with a new and larger telescope, a 50-cm (20-in.) Merz-Repsold
refractor, delivered in 1886. The new telescope joined its companion,
the smaller Merz, in an adjacent, but much larger, cupola on the
roof of Brera (it used to be prominently visible from the botanical
gardens in the courtyard of the Palazzo Brera). As it turned out,
the views of Mercury with the larger telescope were no better than
those with the smaller one – perhaps because the air over Milan
was deteriorating with the growing industrialization of the city,
or perhaps because Schiaparelli's eyesight was no longer as keen
as it had once been.

Meanwhile, poor Terby was becoming more and more impatient,
and implored his friend to publish his beautiful results on Mercury.
At last, on 15 November 1889, Schiaparelli finished his great memoir,
Sulla rotazione e sulla costituzione fisica del pianeta Mercurio (On the
Rotation and Physical Constitution of the Planet Mercury), one of
the classic works of the visual era of planetary observation.[30] It was
well argued, entirely convincing – and, as we now know, almost
completely wrong. With this work, Schiaparelli included his little
planisphere, which, in general effects, resembles his drawing of
20 August 1882, made when Mercury had just passed superior
conjunction.

Schiaparelli observing through the 50-cm (20-in.) Merz-Repsold refracting telescope at Brera Observatory. Despite its greater aperture, this telescope did not significantly outperform the smaller Merz. It was later moved to a location outside Milan, where the lens was shattered in an accident; it no longer exists.

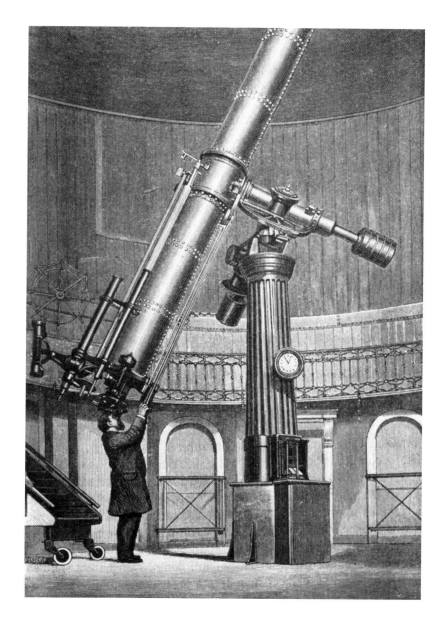

Schiaparelli followed up on 8 December 1889 with an interesting lecture before the Royal Academy of the Lynxes in the Quirinal Palace in Rome, at which King Umberto I and Queen Margherita were in attendance. (This was one of the few occasions on which he travelled outside Milan.) In this lecture he suggested that despite Mercury's synchronous rotation relative to the Sun, it was no 'moon' of the Sun. He had carefully weighed the evidence and had concluded that the planet was surrounded by an appreciable atmosphere. For, he said,

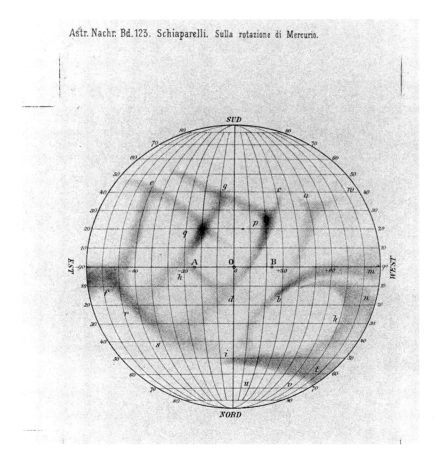

Schiaparelli's influential little planisphere, first published at the end of 1889. The figure-of-5 arrangement appears at the right, and the dark patch 'q' just above and left of centre.

The dark spots of the planet, although permanent in form and arrangement, are not always equally apparent, but are sometimes more intense and sometimes more faint; and it also happens that some of these markings occasionally become entirely invisible. This I cannot attribute to any more obvious cause than to atmospheric condensations similar to our terrestrial clouds.[32]

But the circulation of this atmosphere from the night side onto the day side might well serve to moderate the temperature all around the planet. He explained the nature of the librations, and his reasons for believing in existence of the moderate 'twilight zone' that alternately basked in sun and was sheltered by shade, and left the door slightly ajar to the possibility that the 'wretched star' touched by 'flame' and 'ice' might even be inhabited. Thus, he continued, framing his argument (as he often did) with double negatives:

Considering the difficulty of making a proper study of the dark spots of Mercury, it is not easy to express a well-founded opinion on their nature. They might simply depend upon the diverse material and structure of the solid superficial strata, as we know to be the case with the Moon. But if anyone, taking into account the fact that there exists an atmosphere upon Mercury capable of condensation and perhaps also of precipitation, should hold the opinion that there was something in those dark spots analogous to our seas, I do not think a conclusive argument to the contrary could be advanced.[33]

After throwing out these provocative notions, Schiaparelli abandoned the field to others. From then until his death in 1910, he had nothing more to say about the planet. Secure in the thoroughness of his work and confident of his conclusions, he perhaps thought that there was nothing more to say.

Confirmations

Following the publication of Schiaparelli's results, a number
of astronomers tackled Mercury in the hope of confirming his
findings. The rotation period was, of course, the highest priority.
On it everything that could be inferred about physical conditions
on the planet depended. Denning, having started his study before
Schiaparelli announced his results, stuck with an Earthlike rotation
period of 25 hours. The wealthy amateur astronomer Percival
Lowell, who went from being a traveller and writer about the
Far East to the director of a well-equipped private observatory in
Flagstaff, Arizona, produced a Mercury planisphere in 1896, and
confirmed Schiaparelli's rotation period. (E. M. Antoniadi would
later describe Lowell's map as consisting of a 'spider's web' of
illusory canals.[34]) However the most important work was done in
France. René Jarry-Desloges (1868–1951), a keen planetary observer
who like Lowell was possessed of independent wealth, set up
several private observatories for Solar System studies, of which the
first, established in 1907 at Mont Renard (elevation 1,500 m/4,921
ft), was equipped with a 29-cm (11½-in.) Merz refractor. Jarry-
Desloges hired a number of capable assistants to help him in his
work, notably Georges Fournier (1881–1954), who became his
right-hand man and remained with him from 1907 until 1947, and
Georges' brother Valentin (1884–1978), who worked under Jarry-
Desloges from 1909 to 1914, when he was called up by the French
army to serve on the front. Another noteworthy observer of Mercury
during this period was Georges Bidault de l'Isle (1874–1956), who
also served in the army and was wounded at Verdun (for which he
received the Legion of Honour), and combined a legal career with
astronomical interests. He set up his own private observatory at
L'Isle-sur-Serein, with a 30-cm (12-in.) telescope, and made many
observations of Mercury between 1921 and 1925. Then there was the
artist and astronomer Lucien Rudaux (1874–1947), who had a small

Mercury in 1920. Drawings
by Georges Fournier with
a 37-cm (14½-in.) refractor
at René Jarry-Desloges'
observatory at Sétif, Algeria.
Upper row, left to right:
15 April, CM = 0°; 18 April,
19 April, 23 April, 23 April,
26 April, CM about 54°.
Middle row: 2 May, CM =
82°; 3 May, 3 May, 7 May,
7 May, CM about 105°.
Lower row: 10 May, CM =
118°; 10 May, 12 May,
14 May, CM about 135°.
CM is the Central Meridian.

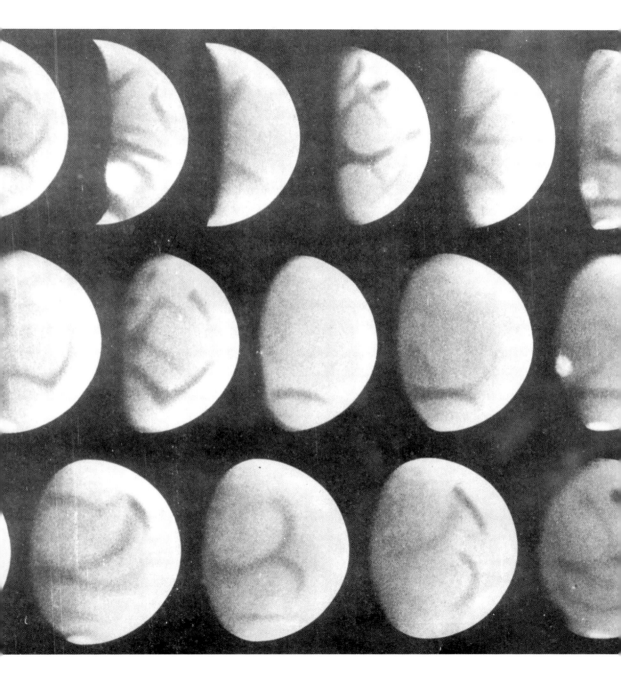

observatory at Donville-les-Bains in Normandy.[35] His observations continued from 1893 to 1927.

All of these observers – and others – produced planispheres of Mercury, while all but one confirmed Schiaparelli's rotation period. That one was Georges Fournier, who could not reconcile his numerous observations from 1907 to 1927 with Schiaparelli's rotation period. Instead he suspected that the actual rotation period was somewhat less than 88 days.

This was the situation when one of the greatest planetary astronomers of the entire visual era decided to enter the field. Eugène Michel Antoniadi, or Eugenos Mihail Andoniadis, as he was christened (1870–1944), had been born in Constantinople (now Istanbul).[36] Though little is known about his background and early life, we do know that by his late teens he was already making his mark as a keen-eyed and judicious observer and an astronomical draftsman of extraordinary skill, despite using only a 75-mm (3-in.) refractor. He made his observations both in Constantinople and on the island of Prinkipio in the Marmara Sea, and by 1891 he had joined the Société Astronomique de France (SAF), founded just four years before by the celebrated astronomer and popularizer of astronomy Camille Flammarion. At about this time he also became one of the original members of the British Astronomical Association (BAA). Probably invited by Flammarion – and sensing, no doubt, the need for more stimulating surroundings than were to be found in the Ottoman Empire – he left Constantinople in 1893 and travelled to France, where he was hired as an assistant to Flammarion at the latter's private observatory at Juvisy-sur-Orge, 18 km (11 mi.) southeast of Paris. Though no doubt an inspiring person to work for, Flammarion expected a great deal from his assistants for a not very substantial sum, and upon marrying in 1902, Antoniadi resigned from Juvisy and the SAF. For a time he seems to have been quite seriously considering emigrating to England, but later changed his mind, and in 1909 established his reputation as a leading planetary observer when Henri Deslandres (1853–1948), the director of the Meudon Observatory near Paris, put at his disposal the 'Grand Lunette', the largest refractor in Europe with a lens of 83 cm (33 in.) fabricated by the Henry Brothers. His observations of Mars that September showed a tremendous amount of natural-looking detail, and effectively demolished Percival Lowell's visions of 'canals' on Mars. In fact Lowell – the man with the tessellated eyeball – drew linear markings (canals, if you will) not only on Mars and Mercury, but on Venus, and even the satellites of Jupiter.[37] He repeated his success with Mars in 1911. His drawings of Mars with the great telescope are marvellous testaments to the observer's skill, and to this day probably remain unsurpassed of their kind.

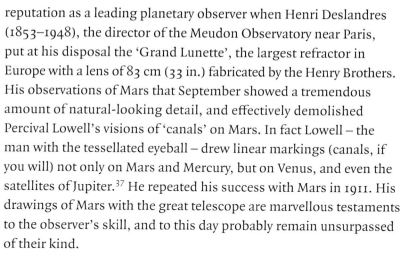

Though Antoniadi had little use for Percival Lowell, whose maps of Mars, Mercury, Venus and so on he believed to be covered in completely 'illusory' details, he was always a great admirer of Schiaparelli. Regarding the Martian 'canals', he discriminated

Camille Flammarion's chateau at Juvisy, with the cupola housing his 23-cm (9-in.) Bardou refractor, as it looked in 2014. Though Flammarion died in 1925, his second wife, Gabrielle Renaudot Flammarion, continued to live here until her death in 1962. The building deteriorated considerably and though it is still partly in ruins, the cupola on top is new, and the refractor has been refurbished and is once more in use by members of the Société Astronomique de France.

between Lowell's illusory markings and Schiaparelli's 'canals', which had at least 'a basis in reality'. Thus he wrote, 'In the positions of each of them, single or double, on the surface of the planet, there is present an irregular trail, a jagged edge of halftone, an isolated lake, in a word, something complex.'[38]

It may have been with something of a predisposing bias in Schiaparelli's favour that Antoniadi – returning to astronomy in 1924 after a break of more than a decade – once more, with the permission of Deslandres, returned to the 'Grand Lunette' and directed it not only at Mars, which came unusually close to the Earth that year, but at

Mercury. He would later call Schiaparelli's discovery of the Mercurian rotation 'the most beautiful of all the telescopic discoveries made by the great Italian astronomer',[39] but at the time he began his systematic study of the planet there were serious doubts. A year before, Edison Petit (1889–1962) and Seth B. Nicholson (1891–1963) had used a thermocouple, a device consisting of two wires of different metals joined together that turns heat into an electric current, on the 1.5-m (60-in.) and 2.52-m (100-in.) reflectors at Mount Wilson to measure temperatures across the tiny disc of Mercury. Their measurements seemed to indicate that the surface temperature on the day side was likely to be about the same as that of the Moon. They also found evidence that some radiation was emitted by the

The 23-cm (9-in.) Bardou refractor at Juvisy, used by Antoniadi in the 1890s. William Sheehan is shown here making some of the first observations with the refurbished instrument – of Mars.

The dome of the great Henry brothers refractor at Meudon. The refractor, with a lens of 83 cm (33 in.), is the largest in Europe. It was used by E.-M. Antoniadi to make his legendary observations of the planets.

dark side as well, 'an indication of a short rotation period'.[40] However, Antoniadi gave short shrift to the physicists' reservations, writing, 'I never accepted the deductions drawn from the interesting indications of the thermo-couple on the planets.'[41] He preferred to place his trust on the tried and true methods of careful visual observations with a large telescope.[42]

He understood the need for prudence since he had not yet made any telescopic observations, but was confident that the 'Grand Lunette', which had performed so well for him on Mars, would help to solve the mystery. He insisted that his research was 'absolutely independent'.[43] In the event, his love for the great Italian astronomer may have had some influence on his results.

Antoniadi did as Schiaparelli – and indeed as all the other important observers of the planet since 1882 had done – and observed in broad daylight. He began his study in the summer of 1924, flushing the planet from the blue background sky with the setting circles of the 'Grand Lunette'. With magnifying powers of 270, 350 and 540x on the instrument, the Mercurian disc never appeared more than twice as big as the Moon as seen with the naked eye. The markings were often quite distinct, and wanting to be absolutely certain of his results, he only drew patches that were definitely seen. He was a persistent opponent of illusions. Instead of the 'crisscrossing bands seen from Milan', Antoniadi found Mercury at Meudon, 'a patchy . . . mixture of dark areas and feeble halftones'.[44] He could not confirm Schiaparelli's report that the spots were pale brown; instead he found them quite as colourless as the dark spots (maria) on the Moon. However, at the very outset of his study, Antoniadi had an excellent view of Schiaparelli's figure-of-5. Thus the 'imprinting' on his imagination began.

Antoniadi tried to observe Mercury when it was near the meridian, and highest in the sky. He followed it for several hours on the same day, and found that the spots – the figure-of-5 – were quite fixed with regard to the terminator. This proved that the rotation period was slow. From one day to the next, they showed a discernible movement that was owing, he thought, to the libration of longitude.

Encouraged by his first successes in 1924, Antoniadi resumed the observations during the summer months of 1927, 1928 and 1929. He could not find the planet near inferior conjunction, when it passed between the Earth and the Sun. But it was 'brilliant' near superior conjunction, and in September 1927, he managed to follow it to within 4° 15' of the solar limb. Under these conditions, the planet showed almost a full disc. Owing to the solar glare, these observations were, he found, extremely tiring to the eye: 'one sees purple after a dozen seconds' observation,' he wrote, and he did not repeat them.[45]

As he continued his study from year to year, Antoniadi found, probably at first to his surprise, that the Mercurian markings were very variable. Sometimes they were 'feeble to the point of being nearly invisible', at other times they stood out 'as boldly as the most prominent features on Mars'.[46] The figure-of-5, whose spots he would name Solitudo Atlantis,

Mercury observed near superior conjunction in 1927. In drawing B (CM = 299° W), the figure-of-5 appears, but in C (CM = 313° W), made three days later, it is only partly visible. Antoniadi regarded such observations as evidence of clouds; in reality, he was unwittingly recording nothing more than the slow-motion drift of the markings due to rotation.

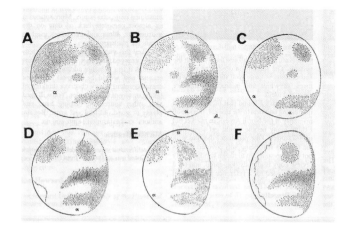

Solitudo Criophori, Solitudo Argiphontae and Solitudo Aphrodites, were apparently particularly affected. Here too he seemed to be following in Schiaparelli's footsteps. As he wrote:

> Schiaparelli had remarked, with wonderful keenness (and it is difficult adequately to admire his work on Mercury) that the aerial veils of that orb are more frequent on the evening than on the morning phase, more frequent over the combination of dusky spots forming his number 5 than on any other region. The writer was enabled to confirm in detail this statement: Solitudo Criophori . . . often rendered invisible by the interposition of local cloud, appeared really much more frequently dimmed by these veils than any other grayish mark, the percentage of cloud over it amounting to about 45, while it curiously stood only as low as 5 over the greatest of the dusky areas,' the Australia-sized dusky patch to which he gave the name Solitudo Hermae Trismigesti [the Wilderness of Hermes the Thrice Greatest] . . . Hence local causes rule very irregularly the distribution of cloud on Mercury.[47]

At the end of his 1929 series of observations, Antoniadi was ready to publish. A first version of his planisphere, without names, appeared that year; another, with names, followed in 1934. So great was Antoniadi's reputation as an observer that his results were to seem the last word for a generation. The doubts of the physicists were effectively routed. In every respect, Antoniadi expressed agreement with Schiaparelli's conclusions: he confirmed the 88-day synchronous rotation period, the libration and the existence of clouds. Since, as we now know, all of these conclusions were wrong, it was a remarkable performance.

Though Antoniadi conceded that it was difficult to conceive of clouds composed of water vapour or ice crystals on such a torrid, sun-baked world, he suggested that the obscurations were presumably not ordinary clouds but veils of fine dust. Yellow dust clouds and even global-circling dust storms were well known to observers of Mars. However, those on Mercury were, Antoniadi claimed, 'whitish, and much more frequent and much more obliterating'.[48]

Later observers who paid attention to Mercury reinforced the seemingly unimpeachable edifice that Antoniadi, building on Schiaparellian foundations, had put in place. Thus the legendary French astronomers Bernard Lyot (1897–1952), Henri Camichel (1907–2003) and Audouin Dollfus (1924–2010) studied Mercury between 1942 and 1950 with the 38-cm (15-in.) and 60-cm (24-in.) refractors of the Pic du Midi Observatory, located at a site 2,877 m (9,439 ft) high in the French Pyrenees, which is famous for exquisite seeing. They found that Mercury's markings did indeed appear to slowly drift from day to day, as had been noted by every astronomer who had ever paid more than a cursory glance at the planet. As usual, this was ascribed to the 'effects of libration and the varying position of the line joining the cusps'.[49] There was no evidence to be found of Antoniadi's dust storms, but there was also no reason to doubt the 88-day rotation. The French team concluded that 'the maps of Schiaparelli and Antoniadi are in good accord with ours if they are rotated about 15 degrees in a clockwise direction. . . . The observations are brought into coincidence by assuming an obliquity of about 7 degrees for the axis of rotation.' Whereas Schiaparelli had claimed that he had established that Mercury's periods of rotation and revolution agreed to one part in a thousand, by 1961 Dollfus had apparently improved on this to one part in

E.-M. Antoniadi's celebrated chart of Mercury, published in 1934. On the right is the figure-of-5 – characteristically seen at the evening elongations – and, as both Schiaparelli and Antoniadi believed, the area most frequently covered with clouds.

The power of suggestion? These planispheres of Mercury were made by a number of independent observers, and yet, with due allowance made for stylistic differences, they show a remarkable consistency from one to another. Top row, left to right: Schiaparelli, 1889; Lowell, 1897; middle row: Jarry-Desloges, 1920; Antoniadi, 1929; bottom row, left to right: markings at the morning elongations, planisphere, and markings at the evening elongations produced from observations by the British Astronomical Association's Mercury and Venus Section, directed by the Glasgow amateur Henry McEwen.

PLATE I. PLANISPHERES OF MERCURY

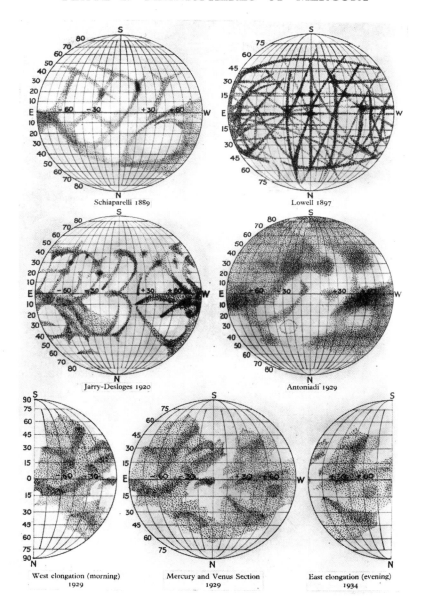

Schiaparelli 1889

Lowell 1897

Jarry-Desloges 1920

Antoniadi 1929

West elongation (morning) 1929

Mercury and Venus Section 1929

East elongation (evening) 1934

The Pic du Midi
Observatory, high
in the French Pyrenees,
whose conditions for
planetary observations
are legendary.

ten thousand. If there was any datum of planetary astronomy that
seemed secure beyond a reasonable doubt, this was it!

At this point, there were very few observers still actively studying
this difficult planet. Notable exceptions included Dale P. Cruikshank
and Alan Binder, who were summer assistants at Yerkes Observatory
in southern Wisconsin during the summers of 1958, 1959 and 1960,
and occasionally had a chance to make daylight observations of
Mercury and Venus with the great 1.02-m (40-in.) refractor. After
Cruikshank and Binder moved to Tucson to start graduate studies
at the University of Arizona, Cruikshank continued to observe the
planet whenever he could, and in February 1962 he produced a
preliminary chart based on the 88-day rotation period. It was to
be one of the last of its kind.

By the early 1960s, as the Space Age was getting under way, there
were signs that something might be seriously off-kilter with regard
to our notions about the innermost planet. As planetary scientist
Clark R. Chapman has said:

We entered the Space Age secure in our portrait of the innermost
planet. Complacency gave way to amorphous uneasiness in
the early 1960s when improved radio-astronomical technology

permitted detection of radio emissions from Mercury . . . The surprisingly strong radiation revealed that Mercury's night side was warmer than the perpetually frigid temperatures expected.[50]

These results were much more difficult to dismiss than the primitive thermocouple results Antoniadi had scoffed at when they showed the same tendency in the early 1920s. Still, the explanation for how heat might be leaking from Mercury's day side to its night side was not obvious to anyone, and in particular, Chapman adds, 'Nobody considered that the supposed eternally dark side might have been basking in the broiling heat of the sun only a couple months earlier.'[51]

The clincher came in 1965. Since 1946, when radio astronomers first used a radio telescope to bounce radio waves off the Moon and receive the reflected signals again, radio astronomers had been ever more venturesome in extending the feat to more distant

The giant radio telescope at Arecibo, Puerto Rico, which in 1965 was used to discover the 58.65-day rotation period of Mercury.

worlds, and Venus in 1957 and then Mercury in 1965 came within range. Using the giant radio telescope at Arecibo, in Puerto Rico, as an emitter and receiver to bounce radio waves off Mercury and receive the reflected signals again, radio astronomers Gordon Pettengill and R. B. Dyce analysed the reflected signal into time-delayed and Doppler-shifted components. The first gave Mercury's distance from the Earth, the second the rate of its spin. The results were surprising, to say the least: they found that instead of the 88-day rotation claimed by Schiaparelli and almost every astronomer afterwards, the rotation period was actually 59 ± 5 days.[52]

A year after this result, Giuseppe ('Bepi') Colombo (1920–1984), professor of Applied Mechanics at the University of Padua and a specialist in celestial mechanics, realized that 59 ± 5 days was consistent with a period of 58.65 days, which is what the rotation period would be if, instead of being locked in 1:1 spin-orbit resonance, Mercury were locked in a 3:2 resonance (that is, one in which it completed three rotations for every two revolutions around the Sun). In showing how this might have arisen, he analysed in detail the tidal torques on Mercury by the Sun. Mercury's orbit is highly eccentric; its distance from the Sun ranges from 46 million km (28½ million mi.) at perihelion to 69.8 million km (43½ million mi.) at aphelion. If, as Colombo suggested, Mercury's globe happened to be deformed into a strongly ellipsoidal shape, the Sun would pull most strongly on its bulge whenever the planet was at aphelion. Eventually whatever rate of spin it had when it formed would be slowed by tidal friction and settle into an orientation where the long axis of its ellipsoid pointed towards the Sun whenever the planet came to perihelion. But there were two solutions, not one: in the 1:1 solution, the bulge would be pointed exactly towards the Sun both at perihelion and at aphelion, while in the 3:2 solution, it would be pointed exactly towards the Sun at perihelion and be at right angles to the Sun at aphelion.

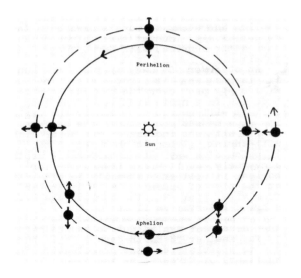

Spin-orbit coupling of Mercury. The planet is considered to have the shape of a deformed ellipsoid with the long axis of the ellipsoid pointing towards the Sun at perihelion and at right angles to the Sun at aphelion.

The 3:2 spin-orbit coupling produces a rather bizarre state of affairs on Mercury. Instead of the same face being perpetually turned towards the Sun, alternate faces of Mercury are pointed towards the Sun each time the planet passes perihelion. The sub-solar points at these times are obviously recipients of the most intense heat and are referred to as 'hot poles'. (One of them has been chosen as the central meridian – 0 degrees of longitude – of the modern coordinate system; the other is at longitude 180 degrees.) The temperature at the hot poles can reach 450 degrees C (840 degrees F), hot enough to melt lead. However, during the night it plummets to minus 170 degrees C (minus 275 degrees F), just above the point at which air would liquefy. Thus the temperature swing is more than 600 degrees C (1,100 degrees F).

A day on Mercury would be rather strange. At one of the hot poles, the Sun would rise in the east, climb into the sky for six weeks, then make a retrograde loop for eight days before continuing westwards and finally setting six weeks later. At longitudes 90 degrees and 270 degrees, the retrograde loop would occur with the Sun at the horizon; thus an observer there would witness double sunrises and sunsets. The retrograde loop occurs because, for eight days near perihelion, Mercury's velocity in its orbit exceeds its rate of spin on its axis.

First Shock . . . Then an Attempt to Understand

Needless to say, the 58.65-day rotation period came as a complete shock to planetary astronomers. Being published in the same summer in which Mariner 4 flew past Mars and sent images showing a heavily cratered surface – also unexpected by most astronomers – it sunk the credit of visual observers to a very low level indeed. It was tempting to suggest that Schiaparelli, Antoniadi and Dollfus, for all their skill and perseverance, had simply beguiled themselves with a fabric of illusion.

Soon after the discovery of the true rotation period of Mercury, Dale P. Cruikshank, who was doing graduate studies at the time at the University of Arizona, and Clark R. Chapman, still an undergraduate at Harvard but accustomed to spending summers in Tucson, attempted to figure out how so many astronomers had been duped. They had a personal interest: both had begun as backyard observers and skilful sketchers of the Moon and planets. According to Cruikshank's recollection,

> Largely at Clark's instigation, we began to look at the Mercury drawings in the *Strolling Astronomer* [the journal of the Association of Lunar and Planetary Observers] and other sources, as well as our own . . . and sitting on the floor in the middle of a big circle of available drawings, we began to piece together a story that might reconcile the observations with the newly determined period.[53]

Before long they hit on the realization that 58.65 days is not only two-thirds of Mercury's period of revolution around the Sun but also almost exactly half of its synodic period, 115.9 days – the period between its successive appearances in the same phase as seen from the Earth. It is also very close to one-third of a terrestrial year (the length of time Earth takes to travel round the Sun).[54] In the interval between synodic periods, Mercury

Dale P. Cruikshank, a leading observer of Mercury in the late 1950s. He made a series of observations of the planet with the 1.02-m (40-in.) Yerkes refractor as a summer student at Yerkes Observatory.

One of the best drawings of Mercury ever made during the visual telescopic era, this composite of sketches by Dale P. Cruikshank combines observations made at Tucson, Arizona, on 16, 18, 19 and 20 January, 1962, with an 11-cm (4-in.) refractor and magnifying powers of 206 and 412x. The CM ranged from 63° to 82° during the five days covered by the sketches.

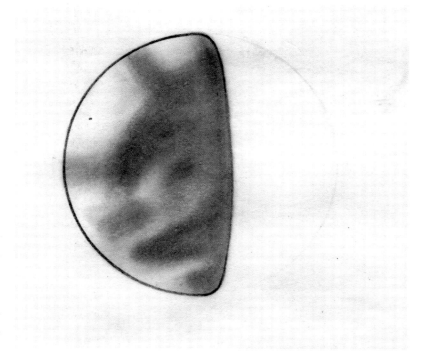

rotates almost twice on its axis, presenting the same face to a telescopic observer. Moreover, as every third evening (eastern) or morning (western) elongation is a particularly favourable one for observers in mid-northern latitudes, observers who selectively view the planet primarily (or exclusively) during these opportune observing windows are presented with more or less the same face of the planet at the same phase for about seven years in succession – a 'stroboscope effect'. Then, because the synchronisms are not quite exact, the periodicity gets out of phase, and observers see for the next several years a region displaced from the first by about 120 degrees, and so on.

These coincidences provide a particularly convincing explanation for the way that observers who kept the planet under scrutiny for only a few years produced drawings consistent with a single map.

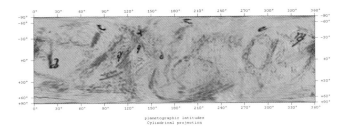

planetographic latitudes
Cylindrical projection

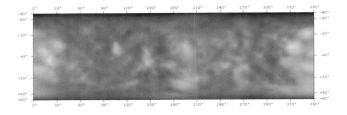

Cylindrical-projection maps by William Sheehan and John Boudreau. The upper map is a cylindrical projection composed of drawings made by Schiaparelli in 1882 and 1883, and the lower on the same scale, is a composite of CCD images by Boudreau. The longitudes increase to the west, the Mariner 10 convention.

However, what about someone like Schiaparelli, who kept the planet under more or less continuous surveillance through seven synodic periods? Chapman notes that even Schiaparelli probably enjoyed the best observing conditions at the favourable elongations, since 'the closer Mercury is to the Sun, the more glare there is in the telescope and from the sky, and Mercury is much smaller'.[55] When Schiaparelli was presented with views of inferior quality, his imagination could have wandered back to the familiar spot arrangements.

Schiaparelli's imagination was also aided by a number of his assumptions. One was the putative librations. Because of the large orbital eccentricity and inclination of the planet, he calculated that Mercury ought to librate 23 degrees in longitude and 7 degrees in latitude. A few observers actually computed the librations; most simply assumed that, whenever there were shifts in surface features, this was the explanation. (No one, however, ever saw the features shift back again.)

The figure-of-5 also played a role. When Schiaparelli first started his serious campaign of observations, in early 1882, this landmark was conspicuously in view. Clearly, it made as strong an impression on him as, say, the Syrtis Major would to a novice observer of Mars. It became his standard of reference, and set an expectation. Later – and even when he was looking at entirely different regions of the planet – he managed with the exercise of a little imagination to convince himself that he was seeing the figure-of-5 (or parts of it) even when there were other, sometimes only vaguely figure-of-5, features in view. He fitted bits and pieces together, eked out suggestions and carried out the kind of figure completion at which the human brain is so adept. As a last resort, he could always invoke

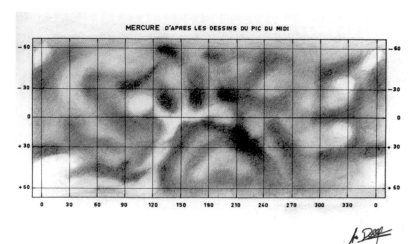

MERCURE D'APRES LES DESSINS DU PIC DU MIDI

After the correct – 58.65-day – rotation period of Mercury was established in the mid-1960s, astronomers re-evaluated the old data by visual observers and produced new albedo maps. These two maps were both published in 1968. The upper one is by Audouin Dollfus, the lower one by Clark R. Chapman.

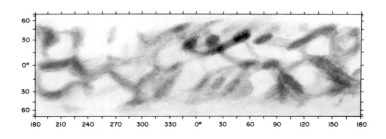

the ultimate fudge factor – obscuring effects such as dust storms. In the end, it was all a neat case of self-hypnosis or autosuggestion.

Clearly Chapman and Cruikshank had caught most of the explanation. But out of curiosity – and motivated in part by admiration for the great Italian astronomer – the present writer teamed up with John Boudreau, a skilful CCD imager of the planets, from Saugus, Massachusetts, to investigate the 'Schiaparelli case' in detail. The first step was to track down all of Schiaparelli's 150 sketches of Mercury from the observing logbooks at Brera, which was done with the help of historian of astronomy Alessandro Manara. The next was to compute modern central meridians to show just what features were on the disc, then to find modern dates corresponding to similar conditions of phase in which these features could be imaged by Boudreau using his 28-cm (11-in.) Schmidt-Cassegrain, and the two compared.[56] It took several

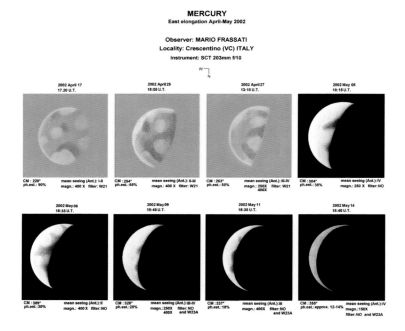

MERCURY
East elongation April-May 2002

Observer: MARIO FRASSATI
Locality: Crescentino (VC) ITALY
Instrument: SCT 203mm f/10

A series of drawings by a skilful visual observer of Mercury, Mario Frassati, at Crescentino, Italy. He used a 20.3-cm (8-in.) Schmidt-Cassegrain to get this series of Mercury observations in 2002, both under daylight and dark-sky conditions.

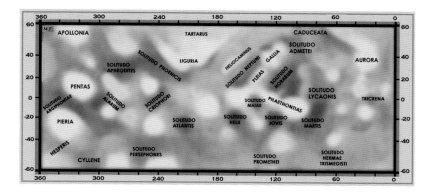

Map of Mercury by Mario Frassati, based on a series of observations from 1997 to 2006. In contrast with Schiaparelli and other astronomers before 1965, Frassati knew the correct rotation period of the planet. Many of the features recorded by earlier observers are readily recognizable here.

years for John to obtain images corresponding to a significant number of the drawings. We were doing what might be called 'historical archaeology'.

It soon became apparent that it was the seeming repetition of the figure-of-5 that threw Schiaparelli off course. He first saw it in early 1882 when viewing the region between longitudes 30 and 90 degrees (as now reckoned). However, during his next observing window, in May and June, he confounded it with another, vaguely similar, feature – lying between 210 and 270 degrees. Thus he entered on a primrose path into error.

His long series of observations shows in detail that 'once a definite expectation is established, it is inevitable that one will see something of what one expects; this reinforces and refines one's expectations in a continuing process until finally one is seeing an exact and detailed – but ultimately fictitious – picture.'[57] Though this particular cautionary tale comes out of planetary astronomy, the same process is inherent in any observations one can make. Were Schiaparelli's rotation period proclaimed as religious dogma, it could never have been challenged; a reminder that only the scientific method – hypothesis generation, checking against data, revising the hypothesis and even abandoning it altogether as data requires it (vagarious as it is, and long as it sometimes takes) points to a way in which errors can be recognized and corrected.

85

FIVE

Mercury Up Close

K nowing the correct rotation period for Mercury meant
that several of the most implausible ideas about the planet
could be disposed of at once. One was the presumption that this
rather moon-like planet had a substantial atmosphere. Another
– though this one was given up with some regret – was the
existence of the balmy 'Twilight Zone'.

With a diameter of only 4,879 km (3,031 mi.), Mercury
is only half as large again as the Moon (and actually smaller
than Jupiter's satellite Ganymede). With a mass only 1/18 that
of the Earth, Mercury's escape velocity – the speed at which a
molecule of gas needs to travel in order to escape its gravity –
is only 4.3 km/s (2¾ mi./s), compared with 11.2 km/s (7 mi./s)
for the Earth. Needless to say, despite the testimony of observers
like Schiaparelli and Antoniadi, it never seemed very likely from
the standpoint of physics that a planet with such weak gravity
would be able to effectively bind gas molecules that were no doubt
violently heated from being in such close proximity to the Sun.
Unless the atmosphere was somehow replenished, it should have
leaked away long ago into space, and how would it have been
replenished on a world likely to be no more geologically active
than the Moon?

We now know that, for practical purposes, Mercury is airless.
Though it does have an extremely tenuous atmosphere, or rather

exosphere, consisting of materials derived from the surface, the pressure is only one one-trillionth that of the Earth (less than about 5×10^{-15} bar). This would meet the standard of a high vacuum were it produced in a laboratory on Earth. Clearly, then, there is nothing to soften the intensity of the solar rays by day or to blunt the edge of bitter cold at night. Mercury must be a desolate place indeed.

The Sun appears greatly swollen in Mercury's sky compared to our view of it from Earth. However, because of the highly eccentric shape of its orbit (eccentricity = 0.206), the Sun's apparent size is some 1½ times larger at perihelion (the point in its orbit closest to the Sun) than at aphelion (the point farthest away), and its track across Mercury's sky is quite as unique as the manner of its rotation. Though relative to the fixed stars Mercury rotates thrice for every two revolutions around the Sun, as seen from the Sun, with respect to which Mercury is tidally locked, it appears to rotate only once every two Mercurian years. Thus from the point of view of an observer on Mercury, one day lasts two Mercurian years. (Remember, alternate faces of Mercury are pointed towards the Sun each time the planet passes perihelion.) Furthermore, as we have seen, because of the eccentricity of its orbit there are 'hot poles' where the temperature extremes are the greatest, as well as, at points 90° of longitude from the 'hot poles', places which perhaps should be called 'least hot poles' which experience double sunrises and sunsets. What a strange world!

Spin-orbit coupling of Mercury. The left diagram is Mercury slightly elongated, showing that the different gravitational forces on Mercury's two ends tend to twist the long axis to point towards the Sun. The middle diagram is Mercury if it had a circular orbit, in which its long axis would always point towards the Sun and Mercury itself would be in synchronous rotation (1-to-1 spin-orbit coupling). The right diagram shows Mercury's elliptical orbit, and how its long axis only points towards the Sun at perihelion – Mercury spins on its axis 1 ½ times during each complete orbit (3-to-2 spin-orbit coupling).

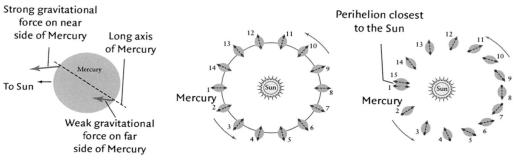

The unusual rotation of Mercury is another byproduct of its eccentric orbit. If its orbit were perfectly or even nearly circular, say like that of Venus (0.007), a 1:1 spin-orbit lock – the situation assumed by Schiaparelli – would have been inevitable. However, we now know that Mercury's orbit undergoes significant changes in shape over millions of years, owing to perturbations by the other planets, especially Venus and the Earth. The orbital eccentricity – currently 0.206 – ranges over time from 0.00, perfectly circular, to 0.45, highly elliptical.

Using supercomputers, Alexandre C. M. Correira of the University of Aveiro, Portugal, and Jacques Laskar of the Paris Observatory have calculated that as the eccentricity of Mercury's orbit increases over time, the probability grows that instead of ending up in a 1:1 spin-orbit resonance it will instead land in a 3:2 resonance. Assuming the planet's eccentricity has changed in this way over the past 4 billion years, they calculated the chance of Mercury locking into a 3:2 state instead of remaining in a 1:1 state. It turns out to be somewhat more than half. Thus the 3:2 spin-orbit resonance in which we now find Mercury turns out to be slightly more probable than the 1:1 case.

What We Learned from Visual Observations

The efforts of visual observers using optical telescopes ended mostly in failure. However, even relying entirely on observations, our knowledge of Mercury would have improved over time. Though the radar observations produced the most dramatic breakthroughs – including not only the rotation period but the discovery of bright patches of water ice in permanently shadowed craters near the planet's poles – the introduction of modern telescopes and techniques such as the charge coupled device (CCD) for imaging would in time have led to significant progress in understanding the planet. (Indeed, amateur-produced CCD images included in this book are far superior to those taken with professional telescopes prior to 1970, and show much more detail than Schiaparelli and Antoniadi had to work with.)

Another significant discovery was of the existence of some of the species, such as sodium atoms, that make up the planet's ultrathin atmosphere (exosphere). Yet another is the planet's comet-like tail, which consists of wisps of exospheric material brushed back by the intense radiation pressure of the Sun. The tail points in an anti-solar direction for a distance of hundreds of times Mercury's diameter. However, it changes in prominence as the planet travels along its elliptical orbit round the Sun. It is most prominent when Mercury experiences the most intense radiation pressure, near perihelion, then fades out as Mercury moves towards aphelion.

The Epic Mission of Mariner 10

Though it cannot be said that no significant progress was made in understanding Mercury from observations from the Earth, the most dramatic jump awaited the development of interplanetary probes. The first probes to the Moon were launched in the late 1950s, and to Venus and Mars in 1962 and 1965, respectively. The first to Jupiter, Pioneer 10, arrived in 1973, with Mariner 10 to Mercury following a year later.

The mission to Mercury had been long in planning. The first step was taken as far back as October 1962, when Michael Minovich, a graduate student at UCLA who had spent several summers at the Jet Propulsion Laboratory (JPL) working on spacecraft trajectories, discovered that in 1970 and in 1973 trajectories to Mercury using gravity assists by Venus would take place. This was just before Mariner 2 – America's first successful interplanetary probe – bypassed Venus (in December 1962). The thought was that the same Atlas/Centaur booster rocket used to send Mariner 2 to Venus could be used to send a Mercury probe as far as Venus, and then

Venus's gravity would help send the spacecraft on to Mercury 'for free'. (To make it to Mercury via a direct route, a much more powerful – and expensive – launch vehicle would have been required, the Titan IIC.)

Direct | Gravity Assist
Mercury
Earth
Titan IIC Launch Vehicle | Atlas/Centaur Launch Vehicle

As usual, it was not as easy as it sounds. As Bruce Murray (1931–2013), Imaging Team Leader for Mariner 10, later recalled, 'There is only one point adjacent to Venus – almost a science fiction "gateway" – that leads directly to Mercury. Each one-mile miss in hitting the invisible moving target translates into a thousand-mile error at Mercury. The voyage to Mercury, economy class, would require unprecedented accuracy in space navigation.'[1]

Trajectories for getting to Mercury. The direct route, left, requires a much more powerful launch vehicle than the Venus gravity-assisted route, right.

Ironically, the mission would receive a much-needed boost from a mistaken Central Intelligence Agency (CIA) assessment, according to which the Soviets, who had been sending probes to Venus every nineteen-month launch window since 1961, might have the same idea as JPL and modify one of their Venus probes for a Mercury mission. Since it was always important to beat the Russians, if possible, this gave a U.S. Mercury mission higher priority. In the event, the Soviets never made the attempt, and the Mercury mission dropped down the list – and in the autumn of 1969 funding was cancelled from the NASA authorization bill in the House. Fortunately, it was later restored in the House–Senate budget compromise, which allowed mission planning and development to proceed.

The spacecraft had to be specially designed to withstand the scorching solar radiation at Mercury – ten times more intense than at the Moon. The Mercury-Venus Mission (or MVM as it was called at the time) required some innovative features to the spacecraft design in order to make it tough enough to survive the harsh

Bruce Murray and friends. In addition to Murray, seated at left, are Carl Sagan (seated, right) and Louis Friedman (standing at left), co-founders of the Planetary Society, and Harry Ashmore, a Pulitzer-prize-winning journalist who was an important advocate for the Society in early years.

environment at Mercury. It was fitted with a gold-plated sunshade to partly shade it from the intense sunlight; also, its solar panels were designed to tilt in order to reduce the solar heating. The main instruments included a charged particle detector to look for radiation belts in the vicinity of Mercury (none were expected), a magnetometer to test whether Mercury has a magnetic field (a test which would only be feasible while the spacecraft was passing over the planet's night side) and twin cameras (which required at least part of the trajectory to pass over the day side). The most critical mission requirements, of course, were for accuracy in aiming and maintaining communications with the spacecraft.

Originally, plans had called only for a single flyby of Mercury. However, that changed dramatically at a February 1970 conference on MVM. The conference was held at Caltech, and Bepi Colombo happened to attend. Murray recounts the tale:

> I barely knew this short, balding man, with one of the most engaging smiles in the world, when he showed up . . . Afterward, he came up to speak to me.
>
> 'Dr Murray, Dr Murray,' he said, 'before I return to Italy, there is something I must ask you. What would be the orbital period of the spacecraft about the Sun after the Mercury encounter? Can the spacecraft be made to come back?'
>
> 'Come back?'
>
> 'Yes, the spacecraft could return to Mercury.'
>
> 'Are you sure?'
>
> 'Why don't you check?'[2]

Colombo, of course, was right. After its first flyby of Mercury,
the spacecraft would be orbiting the Sun with a period of 176 days.
Since this was exactly twice Mercury's period of revolution of 88
days, the spacecraft could be manoeuvred to further encounters with
Mercury every two years. There would be the opportunity to obtain
much more data, though unfortunately at each encounter it would
be presented with the same lighting conditions on the planet.

With an expanded portfolio now calling for not one but three
flybys and a mission lifetime calling for an added year in space,
Mariner 10 (as the spacecraft was officially designated) was sent aloft
in a spectacular night launch on an Atlas-Centaur booster from
Cape Canaveral, Florida, on 3 November 1973. It took advantage
of the second of the two Venus swing-by trajectories calculated by
Minovich long before. Its cameras imaged the north pole of the
Moon as it headed out, and captured a lovely double portrait of the
Earth and Moon. Hitting the sweet spot for the Venus gravity assist
as it passed within 5,670 km (3,520 mi.) of Venus on 5 February
1974, it then headed sunward towards the inner planet, needing
only one more mid-course correction to change its trajectory at
closest encounter from the planet's sunlit side to its dark side (to
optimize conditions for the magnetometer experiment). This was
successfully completed on 16 March. The television cameras had
been switched on twelve hours after launch in order to capture
images of the Earth and Moon on the way out, and again during
the Venus encounter. They were switched on again on 24 March,
as the spacecraft approached to within 5.3 million km (3.3 million
mi.) from Mercury. These first images were no better than those
taken through Earth telescopes (and not as good as the amateur
CCD images included in this book), but at least they showed that
the cameras were still working, and as the spacecraft continued
its approach, more surface details began to emerge. The most
arresting was a prominent bright spot, which would recall to the
more historically informed the brightish spots seen by Denning,

Mariner 10's twin cameras were turned on within twelve hours of launch on 3 November 1973, and several hundred images of the Earth and Moon were captured over the following days. Here, the Earth and Moon appear as imaged from a distance of 2.2 million km (1½ million mi.). The images have been combined in order to illustrate the relative sizes and colours of the two bodies.

Schiaparelli and others. It proved to be a bright-rayed crater, and was given a name – Kuiper, after the legendary planetary astronomer and Mariner 10 imaging team member Gerard P. Kuiper who had died in December 1973 before the spacecraft arrived at Mercury. Not until 1976 did the International Astronomical Union (IAU) approve the official theme of Mercury's craters. At first the IAU Task Group for Mercury nomenclature leaned towards naming the craters after cities or birds. It was Carl Sagan (1934–1996), despite not being directly involved in any way with Mariner 10, who vigorously advocated for naming them after poets and prose writers. Eventually, the IAU agreed and expanded the theme to include all contributors to the humanities and arts – drama, prose, poetry, painting, sculpture, architecture and music. Kuiper was allowed to stay in, since the name was already in wide use; but other prominent craters were given names consistent with the approved theme. Among the prominent craters are Abu-Nuwas, Bach, Beethoven, Cervantes, Chopin, Dostoyevsky, Goya, Hiroshige, Homer, Li Po, Melville, Michelangelo, Proust, Renoir, Rodin, Shakespeare, Sophocles, Mark Twain, Tolstoy, van Gogh, Wagner, Wren, Yeats and Zola.

The closest approach achieved during the first flyby, on 29 March 1974, was only 700 km (435 mi.), and occurred over the night-side of the planet. There was a short period, in fact, when the spacecraft passed behind the planet, and radio contact was temporarily lost. Of course, this was all as planned, and though some of the data was sent back in real time, most of it had to be tape-recorded for later playback to the Earth.

The most important single result concerned the Mercurian magnetic field. At the time it was believed that in order for a planet to generate a significant magnetic field, two criteria had to be met: the planet had to have an iron-rich convecting core to serve as an electrical dynamo, and also a rapid enough rotation to sustain it. Two planets already studied at close range by spacecraft, Venus and Mars, had proved lacking, the first presumably because of its slow rotation and the other because it no longer had a convecting core. The same result had obtained for the Moon, and Mercury was expected to fall neatly into line. Without a magnetic field, the solar wind – the flux of charged particles emanating from the Sun – would stream directly onto the surface instead of being blocked by the magnetic field, and produce a shadow-like cavity behind the planet. As Mariner 10 passed through the anticipated cavity, however, its charged particle experiment detected an unexpected surge of charged particles. The spacecraft had evidently passed through a shock wave, similar to that produced in the Earth's geomagnetic tail.

Though this was strongly suggestive of the existence of a magnetic field, a definitive result was soon obtained by the magnetometer, which showed that Mercury's field was weak – about 1 per cent of the Earth's – but it did indeed exist, and sufficed to explain the bow shock observed in the charged particle experiment. Not only did Mercury have a magnetic field, then, it had a small Van Allen radiation belt as well.

The twin cameras worked flawlessly. For an overview, the approach sequence was used to construct a photomosaic of the incoming side of the planet, and the post-encounter sequence to construct a photomosaic of the outgoing side. Many close-up images provided intriguing views of smaller regions.

'Anyone who sees another planet for the first time,' said Imaging Team Leader Bruce Murray, 'feels a special excitement

Photomosaic of Mariner 10 approach images: the incoming side of the planet.

and sense of wonder. You savour the alien image, the magic of the moment, before you begin systematic analysis. It's like sipping a great hundred-year-old brandy.'[3]

In Mercury's case, the 'first time' brought a strong sense of déjà vu, for Mercury looked strikingly like the Moon. It too had large multi-ring basins, smooth plains recalling the lunar maria, and heavily cratered uplands.

After the first flyby, the spacecraft was hobbled with a number of malfunctions. First, a major electrical short occurred that for a time threatened to dangerously heat up the spacecraft. This was followed by the failure of the tape recorder in August 1974. Real-time television transmission was now the only way to get images from the second encounter, on 21 September 1974. Indeed, the doughty little spacecraft was not through yet; despite its handicaps, it limped along for another two Mercurian years, and was still functioning when it made its third – and closest – encounter, on 16 March 1975, returning high-resolution strips of the planet's surface. The minimum distance was only 327 km (203 mi.). The end came just eight days later, when Mariner 10 ran out of 'attitude-control' fuel, and, no longer able to stabilize itself, began to tumble uselessly out of control. By then it had long since achieved everything hoped for.

All told, Mariner 10 provided high-resolution imagery of about 45 per cent of Mercury's surface, though not all of that was very informative, since some images were taken under nearly high noon conditions; thus the landforms cast no shadows and are difficult to make out.

Smooth plains covered about 40 per cent of the Mercurian surface within the Mariner 10 zone of coverage, with radar mapping from the Earth providing

Photomosaic of Mariner 10 post-encounter images: the outgoing side.

Mariner 10 first discovered the inter-crater plains on Mercury, which are the oldest visible surface units on the planet; they pre-date the heavily cratered terrain, and are typical of much of the planet's surface. This image from MESSENGER shows the lobate scarp Santa Maria Rupes, discovered by Mariner 10, which cuts through both plains and large craters.

supplementary information about
the remainder of the planet, which
appeared to be about 20 per cent smooth
plains. Characteristic features of these
smooth plains are fresh-looking craters
with bright systems of rays. The most
conspicuous is Kuiper, mentioned above;
it is 62 km (38 mi.) across, and is the
brightest feature on the planet. Other
centres of bright ray systems include
Degas, Debussy, Snorri and Copley.

Kuiper, a fresh impact
crater on Mercury with
a diameter of 62 km
(38½ mi.), which closely
resembles Tycho on the
Moon. MESSENGER image
from 2008.

The dominant feature on the surface
of Mercury is the huge multi-ring plain
known as the Caloris basin (the Basin of
Heat), which lies near one of the hot poles of the planet. Mariner
10 captured it right at the terminator – the line dividing day and
night – so that only part of it (the eastern part) was imaged. This
led to an underestimate of its diameter. Based on the Mariner 10
results, the basin was estimated to measure 1,300 km (800 mi.)
from rim to rim, though the later MESSENGER spacecraft, whose
imagery covered the feature in its entirety, has upped this to 1,525
km (948 mi.). In any event, it is clearly colossal, and belongs to the
same class of features as the Imbrium or Orientale basins on the
Moon or the Hellas basin on Mars. So violent was the impact that
formed it that it is believed by many planetary researchers to have
shaped surface features on the other side of the planet – the peculiar
hills and valleys there having, in their view, been sculpted by the
convergence of seismic waves from the Caloris event.

In more detail, the Mariner 10 images showed the usual bull's-
eye structure characteristic of a multi-ring basin. A rim of concentric
massifs (known as the Caloris Montes and Nervo Formation)
surrounds the light-coloured basin interior; the material filling the
latter is thought to have originated either as volcanic flows or as

impact melt. An outer darker rim (the Odin Formation) consists of ejecta deposits from the impact that gives way in turn to radially grooved plains and overlapping craters in clusters formed by larger fragments of ejecta material (the van Eyck Formation). As with Orientale, the best-preserved multi-ring basin on the Moon, the geology of Caloris is extremely complex. Thus, as described by Murchie et al., the interior plains exhibit 'wrinkle ridges' thought to have been formed by a combination of thrust faulting and folding and oriented both concentrically and radially to the basin as in the basalt-filled lunar basins (maria) on the Moon, and troughs. The troughs are graben-depressed blocks bordered by parallel faults – produced by extensional stretching of the Mercurian crust, and in places form what appear to be giant polygon structures.[4]

Caloris basin, one of the largest impact features in the Solar System, measuring 1,550 km (961 mi.) across, shown in a Mariner 10 photomosaic. It was formed early in the planet's history by the impact of a colossal asteroid, 240 km (150 mi.) across. During the Mariner 10 flyby encounters, only half of Caloris basin lay in the sunlit hemisphere of Mercury.

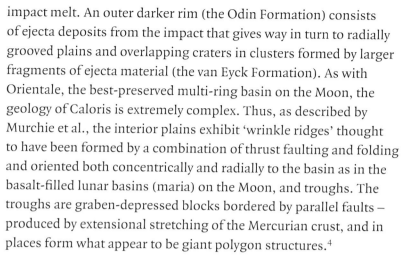

Caloris basin overlaps with one of the 'hot poles', the one at 180° longitude W (or 180° E), which lies just inside the basin's outer rim. The other hot pole lies on the opposite hemisphere, at 0° of longitude. In contrast to the Earth, where the 0° of longitude is precisely defined by the Greenwich Meridian, on Mercury it is longitude 20° W that is precisely defined and forms the Mercurian prime Meridian. By convention, the International Astronomical Union has defined this longitude as running exactly through the position of the small crater 'Hun Kal'. (The name means 'twenty' in an ancient Mayan language.)

Making matters somewhat confusing, though the Mariner 10 team followed the 1970 IAU convention in which longitudes were taken to increase towards the east, the MESSENGER team used the opposite convention, in which they increase to the

west. Thus, to be perfectly unambiguous, coordinates should be explicitly indicated as either degrees W or degrees E. (For instance, the longitude of Hun Kal would be 20°W but 340°E.)

To facilitate comparisons with classical (pre-spacecraft) observations, note that the Mariner 10 zone of coverage lies between longitudes 10° and 190° W. This area includes the 'figure-of-5', consisting of the spots that Antoniadi named Solitudo Martis, Solitudo Lycaeus and Solitudo Horarus. The other hemisphere – including Antoniadi's Solitudo Atlantis, Solitudo Criophiri and Solitudo Phoenicis – lay beyond Mariner 10's terminator. It is fascinating to compare the old observations with Mariner 10's mapped features. It turns out that many real features were detected – though of course the observers did not realize exactly what they were seeing. Denning's 'bright spot surrounded by luminous veins' observed in November 1882 was clearly a glimpse of the young

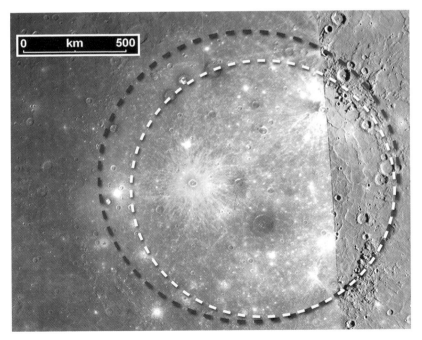

Caloris basin as seen by MESSENGER under high illumination. The white circle is the circumference of the basin as estimated from Mariner 10 images, which showed only the part lying in sunlight; the rest lay hidden in darkness. The black circle is the larger circumference based upon MESSENGER imaging. The true diameter rim to rim is 1,525 km (948 mi.).

impact crater Degas with its extensive network of brilliant rays.[5] Many of Schiaparelli's and Antoniadi's variable bright spots, which they interpreted as clouds, have also been identified as rayed craters.

To return to the Mariner 10 results, the spacecraft also discovered many examples of long sinuous cliffs, called lobate scarps, on Mercury. In the 45 per cent of Mercury's surface imaged by Mariner 10, they cut across all major geological units, and exhibit a broad distribution of orientations. It is thought that Mercury, like the other planets and the Moon, was born 'hot' (due to heat from rapid accretion of planetesimals that provided the raw materials to form the planet and from the decay of early short-lived radionucleotides). As the planet began to cool, its crust contracted, causing the scarps. Though the rate of cooling probably peaked some 2 or 3 billion years ago, it is probably still continuing today. The cooling of the planet would also have affected volcanism on Mercury. On the Moon, which also has undergone cooling since its formation, lava filled the basins within the first half billion years or so after the basins formed, to form the maria that appear as the spots in the 'Man in the Moon'. On Mercury, it seemed likely that the plains represented volcanic flows as well. However, Mariner 10 was not able to determine this, and could not rule out the alternative hypothesis – that they were impact ejecta like the Cayley plains on the Moon.

From the surface of Mercury we turn to the interior, and recall what was perhaps the greatest surprise of the Mariner 10 mission, Mercury's global magnetic field. Weak as it is – recall that its intensity is only 1 per cent of the Earth's – it was not expected to exist at all, since Mercury had a slow rotation and did not seem likely to have a molten iron core, both of which were thought to be necessary prerequisites. The theory that had seemingly worked in the cases of the Earth, Venus, Mars and the Moon broke down in the case of Mercury. Clearly, the magnetic field had something to do with the fact that Mercury has a singularly massive iron core, with a radius (according to the latest estimate) of over 2,000 km (1,200 mi.), equal to 85 per cent of the radius of the entire planet. The other 15 per cent consists of an intermediate mantle of silicates overlain by a thin brittle crust. (On the Earth, the core is relatively small, and the overlying mantle accounts for most of the Earth's radius.)

It has long been known that Mercury is unusually dense. Despite its small size, it has the second-highest density in the Solar System (5.427 g/cm^3), after only the Earth (5.515 g/cm^3). We now know the reason for this: it is an iron planet. It is, in fact, some 60 per cent iron by weight.

To account for Mercury's magnetic field, scientists developed a new model of magnetic field production, positing the existence of currents in the innermost (and presumably still molten) part of the iron core.[6] Whether or not this would prove to be correct was one of the questions that awaited future spacecraft reconnaissance to resolve.

Despite having a magnetic field, Mercury has intense interactions with the solar wind that the magnetic field does only a little to buffer. It also suffers, as any airless body in the Solar System must, from the ongoing onslaught of meteoric impacts. This combined meteoric and solar wind particle bombardment greatly darkens the surface of many airless bodies in the Solar System, but especially Mercury, both because of its closeness to the Sun and the fact that the meteoric bombardment is by interplanetary dust particles at extremely high velocities. The darkening is called 'space weathering'.[7] This generally dark background forms the backdrop against which a few bright craters stand out as radiant centres of brilliant rayed systems. Some of the ray systems are extremely far-flung, reaching thousands of kilometres across the planet. Thus, and in contrast to the Moon, where the lava plains stand out as dark areas against a lighter background, Mercury is a kind of negative image to the Moon – the overall background is dark, and overlain with a smattering of brighter centres. Yet another difference is that on Mercury many large swathes are red high-reflectance plains, and hence of higher albedo than the average: for instance, the Caloris interior plains mentioned above and the expansive Borealis Planitia (Northern Plains).

The bright craters with their systems of rays are obviously relatively fresh, and given the general corrosiveness of Mercury's environment, are evidently younger than similar lunar features – Tycho, for instance. Cratering expert Clark R. Chapman estimates that they are some three times younger. Some of these fresh craters may have been produced by impacts of Sun-grazing comets and periodic comets like Encke – the latter, with the shortest period of any periodic comet at only 3.3 years, passes inside Mercury's orbit and produces a meteoroid stream through which Mercury regularly passes – though it appears that asteroids have totally dominated the entire impact history of Mercury, as also of the Moon and Mars.[8]

Consisting of rock and metal, most asteroids orbit the Sun in the main Asteroid Belt, between Mars and Jupiter. However, a not-negligible number have gone rogue. These include the NEOs (near-Earth objects), which not only have elliptical orbits but generally have smaller semi-major axes – that is, they have

Mosaic of images taken by NASA's Dawn spacecraft shows Vesta, at 525 km (326 mi.) in diameter the second-largest asteroid after Ceres. Note the towering mountain at the south pole (bottom), and the set of three craters that look rather like a snowman (top left).

been moved, on average, closer to the Earth's orbit and the Sun than the main-belt asteroids. There are a few asteroids that orbit entirely within the Earth's orbit, while others – like Icarus – have extremely elongated orbits. Icarus crosses inside the orbit of Mercury at perihelion, but withdraws to the orbit of Saturn at aphelion.

The mention of comets conjures up the possibility that ice from the outer Solar System might even have reached Mercury from time to time, and might remain there still in areas around the planet's poles. Indeed, this possibility received strong support when radio astronomers using the 70-m (230-ft) Goldstone Solar System Radar and Very Large Array (VLA) in the early 1990s detected strong radar reflections from regions near the Mercurian poles (water ice strongly reflects radar). To call these areas polar caps would be to rather overstate the case, but there certainly seemed to be patches, and they were greater and more frequent than those found in the polar areas of the Moon.

To suggest that ice could exist anywhere on such a hellishly hot planet seems paradoxical, and the result was almost entirely unexpected. However, it should be pointed out that Mercury's axial tilt is almost zero; the best measured value is only 0.027° (compared to 23½° for the Earth). This is the smallest value of any planet, with the next smallest axial tilt belonging to Jupiter, at 3.1°. What this means is that an observer situated at either of Mercury's poles would never see the centre of the Sun rise more than 2.1 arc minutes above the horizon, while in some of the deeper craters, the Sun would never rise at all. In these regions of eternal gloom, the temperature would remain frigid, and any ice deposited therein would be able to survive.

Though most astronomers thought that water ice was the best explanation of the radar observations, an alternative theory was sulphurous compounds, and to prove the case definitively one way or the other was among the most pressing questions for the next space probe to Mercury, MESSENGER (the MErcury Surface, Space ENvironment, GEochemistry and Ranging spacecraft). Intended as an orbiting mission rather than a flyby, it was sent aloft aboard a Delta II rocket from Cape Canaveral on 3 August 2004, thirty years after Mariner 10's flyby. To its results we now turn.

MESSENGER *to Mercury*

As with Mariner 10, MESSENGER got to Mercury by means of
a series of gravity assists involving the Earth and Venus, and
incidentally obtained some breathtaking views of the Earth, the
Earth-Moon system and Venus en route. The more complicated
trajectory saved fuel, and in contrast to Mariner 10, which was a
flyby mission, MESSENGER was meant to enter orbit around
Mercury, a much more intricate manoeuvre. Three close passes
by Mercury, in January 2008, October 2008 and September 2009,
braked its motion sufficiently so that at the fourth encounter, in
March 2011, it was able to enter a closed polar orbit around the
planet. During the Primary Mission (first Earth year) each orbit
took about twelve hours; thereafter, for the Extended missions,
the period was shortened to eight hours.

As MESSENGER approached, astronomers were eager to view
those parts of the planet that had been in darkness or poorly
visualized because of high Sun when Mariner 10 flew by – areas
that, despite having been roughly filled in by Earth-based radar
mapping, remained *terra incognita* (or rather *mercurius incognita*).
Over these vast swathes a question mark still hung, and it was
possible to indicate them with the phrase, 'Here be Dragons.'

On the eve of MESSENGER's arrival, Dale P. Cruikshank
recalled some of the emotions he felt when, as a summer student
at Yerkes Observatory in the 1950s, he trained the 102-cm (40-in.)
refractor on the then almost inscrutable little planet.[9]

> I still remember the thrill of bringing little Mercury into the
> field of view, using the setting circles, of course, and after
> some adaptation, seeing a few dusky markings emerge.
> I think most of my Mercury (and some Venus) observations
> were made on lunch hour, with the planets high in the sky.[10]

David Graham of the British Astronomical Association, another long-time Mercury observer, shared with me his musings as history was about to be made. 'MESSENGER,' he wrote, 'will be remembered as the last spacecraft to provide the first close-up views of a planetary surface that had been studied and mapped by telescopic observers during the nineteenth and twentieth centuries. Its mission [thus] marks a watershed. Space probes will continue to survey the Solar System, but never again will one transmit images of

Artist's conception of MESSENGER entering orbit around Mercury.

The launch of MESSENGER, 3 August 2004, aboard a Delta II rocket.

Mercury looming. A MESSENGER high-resolution mosaic of Mercury, captured with the Narrow Angle Camera (NAC) of the Mercury Dual Imaging System (MDIS) on 29 September 2009 as the spacecraft approached the planet for the first flyby of Mercury. Among the details are the Rachmaninoff basin and a bright rimless depression that Imaging Team scientists speculated might be an explosive volcanic vent.

Mercurius Incognita. This image shows an orthographic projection of a global mosaic obtained by MESSENGER. It is centred on the hot pole at 0°N, 0°E, thus showing the hemisphere of Mercury not imaged by Mariner 10. The bright-rayed crater near the bottom of the globe is Debussy.

new lands that invite comparisons with old eyepiece impressions.'[11] In fact, Graham was right. Like Mariner 10 in its time, MESSENGER was a staggering success, and would offer a rather definitive perspective on this little world. It was expected that the mission would end in 2012; however, since the spacecraft continued working well, additional funding was found to support the mission for another three years – an accomplishment that seems little less than astounding given the hostile environment in which it was forced to operate.

Caloris basin revealed. During the three Mariner 10 flybys of Mercury, the same areas were in sunlight and in darkness, and though Caloris basin stood out as obviously one of the most imposing features of the planet, only part of it lay in sunlight and could be visualized, The orbiting MESSENGER was able to show this colossus in its full glory and majesty, as in this mosaic, which has a resolution of 250 m or 820 ft/pixel.

Since the asteroid impact that formed it, it has been filled by red high-reflectance volcanic plains whose colour distinguishes them from the surrounding terrain. Later impacts have excavated low-reflectance material from beneath, and the whole interior has a complex tectonic history.

Sometimes swinging precariously as close as tens of kilometres from Mercury's surface, the polar-orbiting MESSENGER took 250,000 images – some showing the surface at resolutions of only 6 m (20 ft). The entire surface was successfully mapped. (Here be dragons! no more.)

Gravitational perturbations from the Sun altered these orbital paths so that MESSENGER's own propulsion was needed to maintain them. For the last half year or so, the spacecraft would drift towards the planet, then be raised to higher elevations by the spacecraft's propulsion system, then drift down again. When the last bit of fuel was used up, it began to drift down again at roughly the same rate as before, and finally – after completing 4,105 orbits of the planet, on 30 April 2015, it crashed into the raised wall of a crater on the surface (at 54.4° N and 210° E). (In contrast to the 1960s *Ranger* missions, it did not take pictures on the way down; in fact, it crashed at a location on the hemisphere not visible from the Earth.) It is likely that a future spacecraft will one day locate the fresh crater, about 15 m (49 ft) across, formed by MESSENGER on impact.

A wealth of striking features were imaged by MESSENGER, including a rayed crater at 34° S previously known from radar studies, and found in the same position as one of the bright spots occasionally seen from Earth by amateur astronomers. It has been named Debussy. It is 85 km (53 mi.) in diameter, and has a ray system extending 1,900 km (1,180 mi.), which make them as extensive as those of Tycho on the Moon. Another even more impressive rayed crater lies in the far northern latitudes, and was first imaged within ninety minutes of MESSENGER's closest approach during its second flyby on 6 October 2008. Named

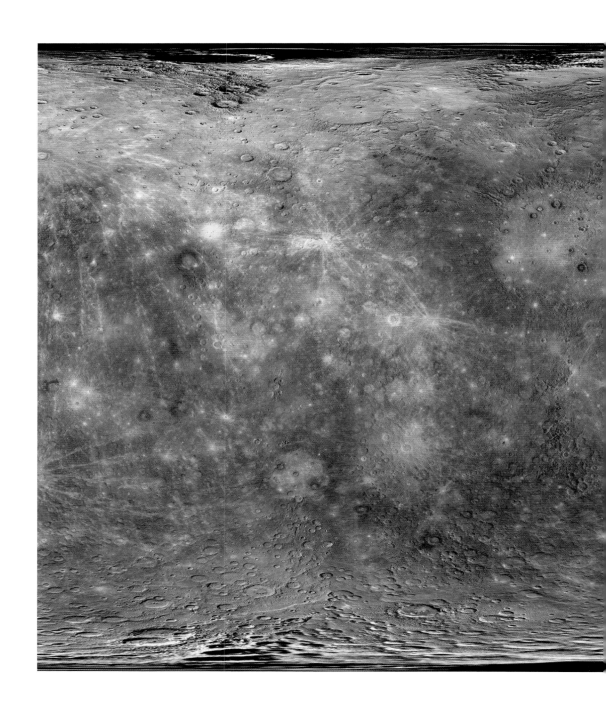

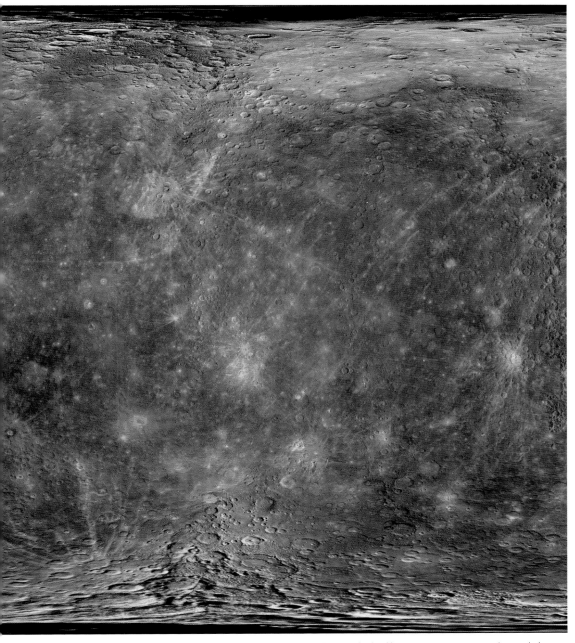

Colour base map of Mercury, based on images obtained by MESSENGER. Lavas are colour-coded orange; low-reflectance materials blue. Note the vast northern volcanic plains, located at the top of the map.

Hokusai, after the Edo printmaker Katsushika Hokusai, the crater lies at 58° N latitude, 18° E longitude, and has an extensive network of bright rays reaching at least 4,500 km (2,800 mi.), thus extending all the way into the southern hemisphere. The rays are thought to be produced as rocks and boulders thrown out by the impact dig below the darkened surface and expose brighter materials below,

but the discovery of rays of such length on Mercury came as a distinct surprise. Because of Mercury's higher surface gravity, its ejecta (including the materials that form rays by turning over brighter material) ought to be tighter and less far-flung than those of the Moon. A possible explanation for the discrepancy is that the craters Debussy and Hokusai are significantly younger than Tycho (whose age was estimated from Apollo 16 lunar samples as 109 million years), and thus have not yet suffered as much degradation from space weathering – no doubt much more efficient on Mercury than on the Moon. However, the most probable explanation is that impacts – whether from asteroids or comets – average much higher velocities than those on the Moon (though they spread out over a wide range, of course, on average the velocities at Mercury are in the order of 40 km (25 mi.)/sec., more than double those on the Moon). Since kinetic energy is proportionate to the velocity squared, the impacts will produce much wider splatter.[12]

It seems that the existence in the northern hemisphere of several bright-rayed craters and the vast reddish and bright-albedo Borealis Planitia account for the fact that many telescopic observers over the years saw the northern cusp as decidedly brighter than its southern counterpart. (This, by the way, probably provides the explanation for Schröter's blunted southern cusp: it is not, it seems, so much that the southern cusp is blunted as that it is muted in comparison to the polar region on the opposite side.) The illusion of a bright northern cusp or 'polar cap' is especially pronounced when Hokusai is close to the central meridian. Another important MESSENGER finding involves the lobate scarps, which were so prominent in the Mariner 10 images. Would they be equally in evidence over the other 55 per cent of the planet not imaged by Mariner 10? They were. Indeed, they are the most prominent large-scale (tectonic) landform on Mercury, with some reaching lengths of 600 km (370 mi.).

Recall that the Earth's crust is made up of assorted tectonic plates that interact to produce mountains, rift valleys, deep ocean

Enterpise Rupes, about 1,000 km (620 mi.) long and with over 3 km (2 mi.) of vertical scale, is the largest lobate thrust fault scarp on Mercury. A global network of these tectonic landforms were discovered by MESSENGER, and are the surface expression of thrust faults due to the contraction of Mercury.

trenches and earthquakes. By contrast, Mercury's crust is one solid plate. As Mercury's interior cooled and its globe contracted, there were no plates to move around and relieve stress. Something had to 'give', so that the crust could readjust to the global shrinkage, which produced a 7-km (4-mi.) reduction in the planet's radius overall. The visible scarps on Mercury suggest that this contraction was gradual and occurred during a lengthy epoch following the end of the era of heavy bombardment in which the heavily cratered 'uplands' of Mercury's surface formed. (Had the contraction occurred earlier than the era of heavy bombardment, we might not recognize lobate scarp topography so easily.) After the heavy bombardment ended, about 3.8 billion years ago, lava

Victoria Rupes, shown here, is a scarp created by the global shrinkage of Mercury as it cooled. This MESSENGER mosaic spans 250 km (155 mi.).

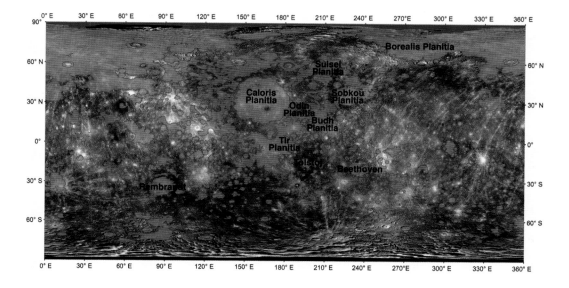

flowed freely onto the surface, filling a wide ring around Caloris basin, which presumably formed about 3.8 billion years ago (the age of the very similar Orientale on the Moon), and covering the vast Borealis Planitia (Northern Plains), which seem to be of about the same age. But then the lava flows seem to have rather abruptly stopped, as the global contraction continued. Indeed, by about 3.5 billion years ago, all of the conduits into deep lava were cut off, and lava flows ended almost completely at that time. (On the Moon, lava flows continued, though at a pretty minimal rate, for another billion years.) Though there are more recent lava flows on Mercury, the only ones so far verified are within the 306-km- (190-mi.) wide basin Rachmaninoff, the bottom of which is the lowest elevation location on Mercury. Presumably the impact that formed this small basin punctured through the closed-off crust to tap into deep magma and created the special circumstances allowing the lava to flow once more onto the surface.

MESSENGER's images show many craters that have been filled partially by lava, and especially on the Borealis Planitia, 'ghost

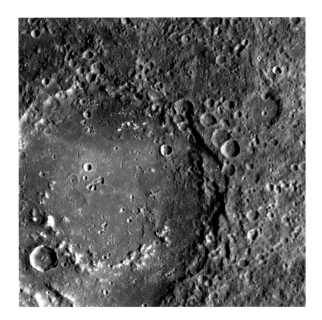

This colour image from MESSENGER, obtained with the Wide Angle Camera (WAC) of the Mercury Dual Imaging System (MDIS) on 13 January 2012, shows Rachmaninoff basin at an image scale of 304 m (397 ft)/pixel. The inner ring of Rachmaninoff's double-ring structure appears blue, showing evidence of Low Reflectance Material, which contrasts with the lighter lavas of the inner ring floor.

craters', like those on the Moon, in which the barest outlines of lava-submerged craters are visible. MESSENGER images also show volcanoes – though they are not noteworthy for oozing lava. Instead they represent places where explosive volcanism has distributed ash (reddish in colour) across the surrounding terrain. Thus they may be more like the dark spots in Alphonsus on the Moon. Many of them are located around the rim of Caloris basin.

Also found in the MESSENGER images was a new kind of shallow, irregular landform, called hollows. These are depressions thought to be caused by the sublimation – passage directly from solid to gas – of volatile substances, such as sulphur (a 'volatile' in a planetary science context, which means an element that evaporates at moderately high temperatures). Once the volatiles have sublimated, the material left collapses. Hollows range from tens of metres to several kilometres in diameter. They are uniquely Mercurian landforms, especially associated with some craters – for instance, Kertész crater, 31 km (19 mi.) in diameter.

The discovery that Mercury is volatile-rich was one of the biggest surprises of the MESSENGER mission. It was simply not expected for a planet presumed to have formed so near the Sun and to be, as Percival Lowell once put it, 'the bleached bones of a world'. Indeed, something like 4 per cent of Mercury's surface material is made up of sulphur compounds. This finding clearly has a significant bearing on theories of the formation of the planets in the inner Solar System.

The crust of Mercury is darker than that of any other known planet or satellite, including its nearest counterpart, the Moon. The albedo – the ratio of reflected light to incident light – is only 7 per cent.[13] (Remember that the photometric work of Zöllner already established this in the nineteenth century, but no one paid any attention at the time.) Iron-bearing minerals, in combination

A mosaic of MESSENGER images obtained with the Mercury Dual Imaging System (MDIS) showing the interior of Abedin crater, 72 km (45 mi.) in diameter (centred at latitude 61.7° N and longitude 349° E). The crater floor is covered with once-molten rock melted by the impact, and cracks that formed as the melt cooled. The shallow depression that lies among the complex central peaks of the crater is surrounded by reddish material, and may have resulted in explosive volcanism at this location.

A close-up view of the expansive smooth plains in the northern hemisphere of Mercury. These plains were created when huge amounts of volcanic material flooded the surface in the past, burying craters and leaving only traces of their rims visible. The large 'ghost crater' in the centre is approximately 103 km (64 mi.) across. Note that the well-preserved craters superimposed on the smooth plains had to have formed after the flooding ended.

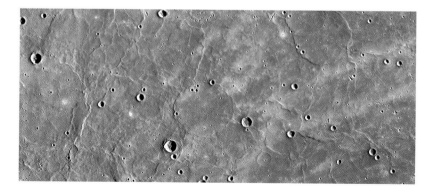

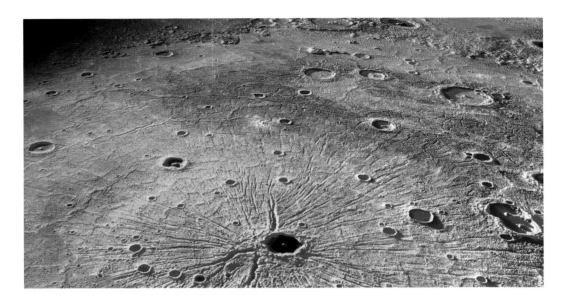

with space-weathering processes such as those described above, were thought to be the darkening agent, as in the case of the Moon. However, MESSENGER found that, abundant as iron is in Mercury's core, it is depleted in Mercury's crust. The explanation for this rather odd finding is thought to be that when Mercury was entirely molten, all the iron sank to the core, while carbon in the form of graphite – the stuff used in the old-fashioned writing utensils known as pencils – on account of its greater buoyancy rose to the surface of the global magma ocean. It is this graphite that appears to be the material responsible for Mercury's low albedo. Not all of Mercury's complement of carbon was necessarily present at Mercury's birth. Rather, new carbon may be regularly transported to Mercury by comets and asteroids.

As stated above, in telescopic views the bright north polar region containing Hokusai crater and the Borealis Planita can sometimes vaguely resemble one of the polar ice caps of the Earth and Mars. Despite the suggestive appearance, ice was certainly not to be expected on a planet so near to the Sun as Mercury.

A perspective view based on MESSENGER Mercury Dual Imaging System (MDIS) and Mercury Laser Altimeter (MLA) data looking northwest over Caloris basin. The mountain range forming the basin's rim makes up the arc in the background, while in the foreground, a set of tectonic troughs – known as Pantheon Fossae – radiates from the centre of the basin outward towards the basin's edge. Apollodorus is the impact crater superposed just off-centre of Pantheon Fossae. White and red code for high topography, greens and blues for low topography; the total vertical differential is 4 km (2½ mi.).

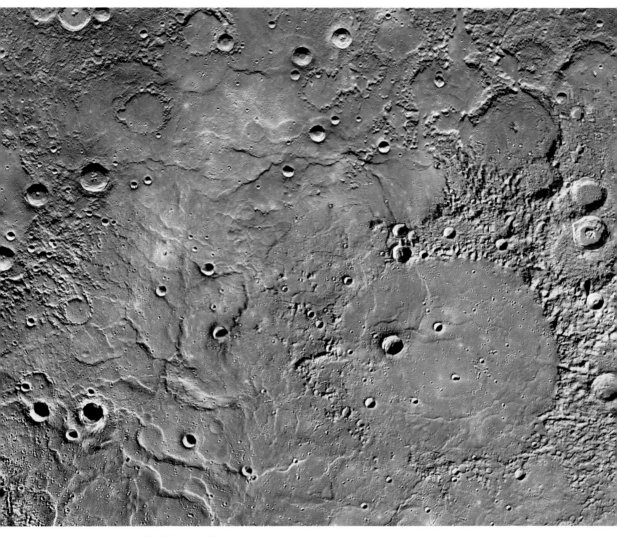

Mercury's northern volcanic plains, shown in enhanced colour to emphasize rocks of different composition. The large impact basin, Mendelssohn, in the lower-right part of this image is 291 km (181 mi.) in diameter. The characteristic landforms in the bottom left are wrinkle ridges, similar to those found in the lunar maria, which formed as lava cooled, and several good examples of lava-filled craters ('ghost' craters) are visible.

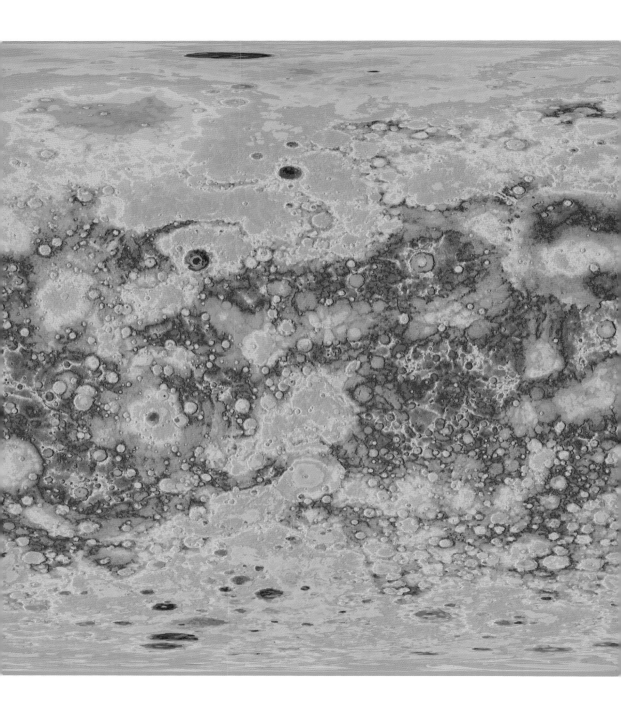

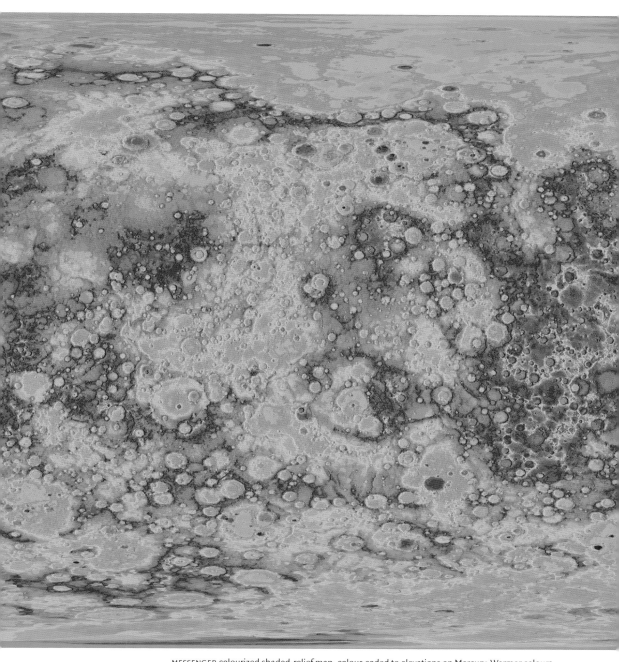

MESSENGER colourized shaded-relief map, colour-coded to elevations on Mercury. Warmer colours depict higher elevations, cooler colours lower elevations.

Nevertheless, in the 1990s radar observations showed highly reflective patches in the polar regions that seemed, in fact, to consist of water ice, though at the time sulphurous compounds were also deemed a possibility. MESSENGER put all doubts to rest. Its detection of hydrogen ions near polar latitudes and detailed images within the shadows, combined with sophisticated thermal model calculations, have established beyond any doubt that it is water ice, not sulphurous compounds, that fill the crater floors. The patches are found in the bottoms of certain craters, such as Prokofiev and Kandinsky in the north, and Chao-Meng Fu in the south – wells of permanent darkness. (Strangely, some of the buried ice patches are actually covered with darker material than the average Mercurian landscapes.)

The temperatures in these isolated pockets are low enough – constantly below -93°C (-136°F) – that water ice can remain stable for billions of years. Likely sources of the water ice are comets and meteoroids containing hydrated minerals, though it may also be that some water outgasses from the planet's interior or is created when solar wind protons interact with oxygen in minerals on the surface. Wherever it came from, polar water ice certainly exists, and indeed MESSENGER found one-third more ice than expected.

Raditladi, a well-preserved peak ring basin on Mercury, has a diameter of 258 km (160 mi.). It was discovered during MESSENGER's first Mercury flyby in 2008, and exhibits well-preserved structure, a relatively young age, exterior-impact melt ponds, a peak ring covered with hollows, and concentric troughs on its floor. This relief map based on MESSENGER images looks towards the north pole of the planet and shows the broad area of the smooth volcanic northern plains. Higher elevations are coded brown and lower elevations blue.

This MESSENGER mosaic is composed of three Narrow Angle Camera (NAC) frames taken on 11 January 2013, and shows the 31-km (19-mi.) diameter Kertész crater (at latitude 27°N and longitude 146°E). Kertész crater is remarkable for its extensive system of hollows.

Hokusai crater, with a diameter of 114 km (71 mi.), is notable for its widely distributed system of rays, which extend across much of the planet. This image shows the interior structure: a ring of central peaks, partly buried within the frozen sea of impact melt filling the crater floor, and a superlative series of terraces marking the rim.

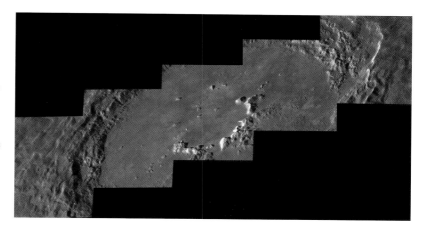

One of the most dramatic of all MESSENGER views, showing a feature nicknamed by the MESSENGER team 'the spider', consisting of the 42-km- (26-mi.-) diameter impact crater Apollodorus surrounded by the radiating troughs of Pantheon Fossae. This feature is located near the centre of Caloris basin.

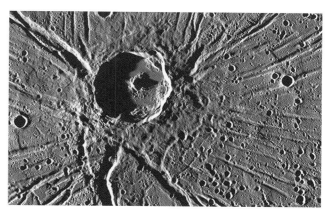

A high-resolution view showing hollows on the southwestern peak of Scarlatti basin. Note the abundance of small impact craters on the surface surrounding the hollows but the paucity of such features on the hollows themselves. This distribution shows the hollows must be very young compared to the rest of Mercury's surface.

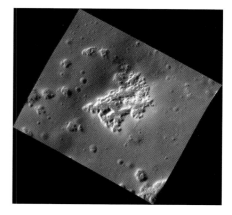

This relief map based on MESSENGER images looks towards the north pole of the planet and shows the broad area of the smooth volcanic northern plains. Higher elevations are coded brown and lower elevations blue.

In the (unlikely) event that a future human mission is sent to Mercury, or a Mercurian base is established, the existence of water ice in the polar regions will undoubtedly make them among the most appealing choices of targets.

MESSENGER added additional definition to the magnetic field discovered by Mariner 10 and shown to be a mere 1 per cent as strong as the Earth's. It is, in fact, so weak that it barely stands off the solar wind from the subsolar point to form the magnetosphere. The magnetic field is produced in the molten outer core, and acts as though it is centred upon a point 480 km (300 mi.) – about 20 per cent of Mercury's radius – north of the planet's centre. This eccentric location leaves the south polar regions more exposed to the solar wind than the north polar regions. The solar wind interactions produce the exosphere of the planet, in which (in contrast to an atmosphere in the usual sense) the constituent particles and molecules travel on collisionless trajectories and are far more likely to strike the surface or escape from Mercury altogether than to interact with each other. Constituents of the

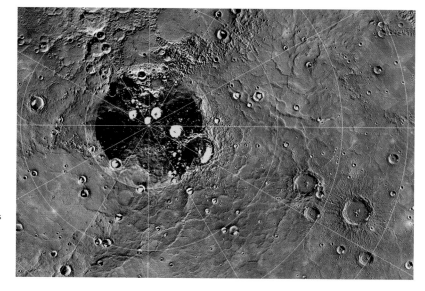

Area around Mercury's north pole. Ice deposits imaged by Earth-based radar are shown in yellow. MESSENGER data shows that all of the polar deposits imaged by Earth-based radar are located in areas of persistent shadow, such as the craters Kandinsky and Prokofiev.

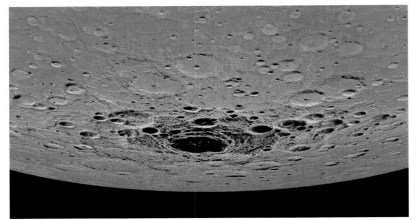

Mercury's south pole. This orthographic projection, based on data from the Mercury Dual Imaging System (MDIS), is colour coded on the basis of the percentage of time that a given area receives sunlight. The largest permanently shadowed region near the centre is the 1,890-km-(112-mi.-) wide crater Chao Meng-Fu, which harbours considerable amounts of water ice.

exosphere include hydrogen and helium, discovered by Mariner 10, sodium, potassium and calcium, discovered in ground-based observations, and magnesium, added by MESSENGER. Neutral atoms – that is, atoms without net electric charge – are accelerated by radiation pressure from the Sun to form Mercury's extended anti-sunward comet-like tail.[14]

Unknown at the time of Mariner 10, Mercury has a liquid iron outer core. It is motions of this electrically conducting medium, stirred by Mercury's rotation, that create its magnetic field. This is another distinction for Mercury, as it is the only one of the inner planets (besides the Earth) to have an internally generated magnetic field.

With MESSENGER's mission ended and its impact-pulverized remains scattered about on the surface, what is the next step for the exploration of Mercury? The European Space Agency and the Japanese Aerospace Agency are currently at work on a pair of Mercury probes. One is to study the planet itself, while the other investigates the magnetosphere. The mission, named BepiColombo in honour of the great Italian mathematician and engineer who first explained Mercury's spin-orbit coupling and showed how to get a spacecraft into a resonance orbit with Mercury to achieve multiple flybys, is scheduled for launch in 2018 and is due to arrive at Mercury in 2025.

There are undoubtedly still surprises in store for us, but at least when it arrives we will know far more about Mercury than when MESSENGER entered orbit in 2008 or when Mariner 10 made its three flybys back in 1974–5.

SIX

VULCAN

Astronomers concerned with calculating the motions of the planets were increasingly perplexed, in the last part of the eighteenth and first part of the nineteenth centuries, by the seeming resistance of the innermost planet to conform to their predictions. This was exposed most clearly by the errors in their calculations of the times of Mercury's transits across the Sun. The prediction of the transit of 1707 had been off by a day; that based on Edmond Halley's tables for the transit of 1753 was off by many hours, while Joseph-Jérôme Lefrançois de Lalande's prediction for the 1786 transit was off by 53 minutes – so that all the astronomers of Paris except one, Lalande's student Jean-Baptiste Joseph Delambre, had given up, packed up their telescopes and gone home – and missed the transit altogether.

Le Verrier Enters the Fray

Astronomy is supposed to be an exact science, and so this state of affairs threw down the gauntlet to the mathematical astronomers whose business it was to keep the planets' motions moving according to train-table precision. After the 1786 debacle, Lalande published revised tables for the transits of 1789, 1799 and 1802. They were revised in their turn by the Thuringian polymath and statesman Bernhard August von Lindenau (1779–1854). Hardly remembered today, Lindenau was a remarkable figure in his time

Urbain-Jean-Joseph Le Verrier, as he appears in the marble statue by Henri-Michel Antoine Chapu, erected in 1889, which stands before the entrance of the Paris Observatory. When he died in 1877, he thought he had discovered two new planets, Neptune and Vulcan.

– not only an accomplished astronomer but a lawyer, influential minister of the interior in Saxony, and art collector whose collection included a Lippi, a Botticelli and a Fra Angelico, all given, at his death in 1854, to found the Lindenau Museum in his native town of Altenburg. (It still exists.) Lindenau's tables were used to make predictions for the 1822 and 1832 transits.

Enter at this point Urbain-Jean-Joseph Le Verrier (1811–1877), one of the greatest mathematical astronomers of all time.[1] Born in St Lô in Normandy in 1811, he began as a skilful chemist but, when his advancement was blocked by the bestowing of a choice position on another, he switched, effortlessly, to astronomy, and soon established his reputation as a specialist of celestial mechanics – the art of analysing the mutual perturbations of the planets. As such, he was assigned in 1841 by François Arago (1786–1853), the director of the Paris Observatory, to take up as his first priority the motion of the singular little planet that had baffled astronomers going back to Ptolemy. (Recall from Chapter Two that Ptolemy had encountered so much difficulty with this planet that he had needed to improvise a unique double perigee solution, thereby departing from the theory that had seemed to work satisfactorily for all the other planets.)

Le Verrier's investigation of Mercury lasted three years, and led to a paper, 'Détermination nouvelle de l'orbite de Mercure et de ses perturbations',[2] which was a decided advance on Lindenau's work. He followed up, two years later, with a book, *Théorie du mouvement de Mercure*,[3] in which he presented new tables for the planet and supplied a prediction for the transit of 8 May 1845, which would be partially visible from France and entirely from North America. It was eagerly awaited by Ormsby Macknight Mitchel (1809–1862), an astronomer at the Cincinnati Observatory. Mitchel himself had acquired the observatory's fine 28-cm (11-in.) Merz refractor; it was such a glass, he declared, 'in search of which I had traversed ocean and land'. As the transit approached, Mitchel's excitement rose to fever pitch. He wrote,

I had the high satisfaction of seeing mounted one of the largest and most perfect instruments in the world. I had arranged and adjusted its complex machinery, had computed the exact point on the Sun's disc where this planet ought to make its first contact, had determined the instant of contact by the old tables [of Lindenau], and by the new ones of Le Verrier, and, with feelings which must be experienced to be realized, I took my post at the telescope to watch the coming of the expected planet. After waiting what seemed almost an age . . . I caught the dark break which the black body of the planet made on the bright disc of the Sun. 'Now!' I exclaimed; and, within 16 seconds of the computed time, did the planet touch the solar disc, at the precise point at which theory had indicated the first contact would occur.[4]

What seemed to Mitchel almost a miracle of prediction, alas, was shattering to Le Verrier. Like most practitioners of his exacting art, he was an obsessive perfectionist, and regarded accuracy good to only sixteen seconds as failure. He was so disappointed, in fact, that he immediately withdrew his tables, then being typeset by the Bureau of Longitudes, from publication, and put Mercury on the back burner while he turned to other pressing problems, including the calculation of comet orbits and the problem of another errant planet, the then-outermost, Uranus. The latter would bring astronomical immortality, as his calculations showed the discrepancy in Uranus's motion was probably due to the disturbing action of an outer planet, and led to the discovery of Neptune in September 1846. Buoyed with this success, Le Verrier in 1849 returned to Mercury. This marked the beginning of a ten-year siege, at the end of which he finally consented to publish what he regarded as a solution to the planet's misbehaviour.

In 1859, the year of Darwin's *On the Origin of Species* and Gustav Kirchhoff and Robert Bunsen's three laws of spectroscopy,

Le Verrier released his long-cogitated and reworked memoir on Mercury to the press. The result, in a word, was shocking. Even after taking everything into account, Le Verrier declared, Mercury still failed to behave as it ought to do according to the irreproachable Newtonian law of gravitation. A small discrepancy stubbornly and exasperatingly clung to the result. It was due neither to observational errors nor to incorrect terms in the perturbation theory.

What he had discovered was that, even after the disturbing effects of all the other planets had been accounted for, of which the largest terms were due to Venus and the Earth, the perihelion of Mercury's elliptical orbit was precessing around the Sun at a rate slightly faster than that predicted on the basis of the Newtonian inverse-square law of gravitation. The amount was exquisitely small, at only 38″ of arc per century. But Le Verrier had no doubt that it was real. Following the same line of reasoning that had led him to the discovery of Neptune, Le Verrier posited the existence of a small amount of additional mass inside the orbit of Mercury.[5] This extra mass did not need to be a planet, but it did need to be present in sufficient quantity to account for the excess motion of the perihelion of Mercury while not affecting other things, such as the motions of Venus or the Earth, where there was nothing amiss.

Supposing it to be a planet, and assuming a mean distance from the Sun of 0.17 AU, where 1 AU = 149,560,000 km (92,956,000 mi.), Le Verrier found that in order to produce the observed effect on Mercury's perihelion, its mass would have to be similar to that of Mercury itself. Such a planet's greatest elongation from the Sun would be only 10 degrees, but it would shine more brilliantly than Mercury, since the brightness of a planet increases with proximity to the Sun. Le Verrier, naturally, asked how such a planet, 'extremely bright and always near the Sun, could fail to have been recognized

Precession of Mercury's perihelion.

The total solar eclipse of 11 July 1991, as photographed from La Paz, Baja, Mexico, where totality occurred at close to noon with the Sun almost overhead. The duration of this eclipse was almost as great as it can ever be – 6 min., 53 sec. at the point of maximum totality – and there will not be a longer total eclipse until 13 June 2132.

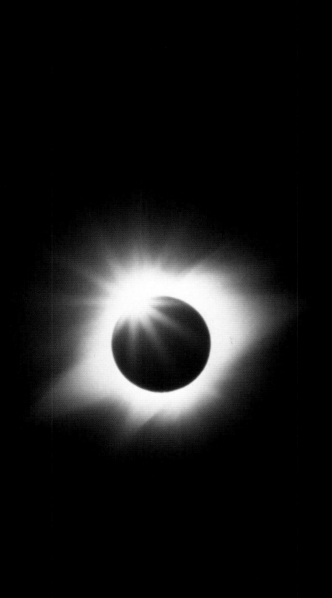

during a total eclipse?' And he added: 'Would not such a planet pass in transit between the Sun and Earth, and thereby make its presence known?'[6]

At that moment, Le Verrier inclined to the hypothesis that the mass disturbing Mercury's motion was not a planet but existed as an intra-Mercurian ring of small bodies, an inner asteroid belt, in other words. Nevertheless, the planet hypothesis lingered in many minds. When, on 12 September 1859, Le Verrier published his conjecture, in a letter to his colleague at the Paris Observatory, Hervé Faye (1814–1902), he added an exhortation to astronomers to keep a close watch on the disc of the Sun and to scrutinize even the tiniest sunspot, in case it might be an intra-Mercurial planet in transit. Faye pointed out that a better opportunity to discover the intra-Mercurial game might occur at the forthcoming eclipse of 16 July 1860, at which the path of totality would cross Spain and thus be accessible to European astronomers. Clearly, a mood of expectation hung in the air.

Like a Bolt Out of the Blue

Unexpectedly, just before New Year's Day 1860, a doctor and amateur astronomer of Orgères-en-Beauce named Edmond Lescarbault (1814–1894) penned a letter to Le Verrier, in which he claimed to have already observed just such a moving spot as Le Verrier had predicted, from his small private observatory, the previous 26 March. It never was entirely clear why he had waited so long to communicate the observation; nor was it obvious why he should only have overcome his hesitation after reading an article in *Cosmos*, a journal edited by the well-known popularizer of astronomy the Abbé François Moigno, which summarized Le Verrier's calculations. It may have been nothing more than that Lescarbault was, as a contemporary described him, a bit of a dreamer. Whatever the case, his interest did not appear to be a recent enthusiasm. He claimed to have been keeping a close eye on the Sun, specifically for the purpose of recording a small transiting planet, did such exist, ever since 1837.[7]

Le Verrier, in receiving such a letter from a complete unknown, was initially, of course, filled with scepticism. At the same time he was intrigued, and so decided to set out immediately from Paris to the small village in order to investigate the matter. As soon as he arrived, he knocked imperiously at the doctor's door. The interview must have been tense, but somehow by the end of it the humble doctor managed to convince the great man of the soundness of his methods and of his personal integrity, for Le Verrier departed that day for Paris convinced – as he would remain until the end of his life – that the intra-Mercurian planet had been seen. He calculated an orbit for it, and even gave it a name, 'Vulcan'.

In fact, Lescarbault's 'Vulcan' was never seen again. To this day what he might have seen remains a mystery. The French popularizer of astronomy, Camille Flammarion, an enemy of Le Verrier because of the way he had been treated by the great man when he had served as a teenaged human computer in the Paris Observatory, attempted to explain the observation away as an error of the crudest kind. He had, according to Flammarion, merely observed a sunspot, and on interrupting his observations, become confused about the orientation of the Sun in his telescope. Another astronomer hostile to Le Verrier, Emmanuel Liais (1826–1900), professed to have been observing the Sun from Brazil at the exact moment Lescarbault made his observation, but saw nothing out of the ordinary. Nothing is certain in this story.[8]

Vulcan's Last Stand

Meanwhile, the 1860 eclipse in Spain came and went, with Vulcan failing to put in an appearance. The 'Great Indian Eclipse' of 18 August 1868, and the North American eclipse of 7–9 August 1869, though rich in astrophysical discoveries, also failed to turn up the much anticipated planet.

What might be regarded as 'Vulcan's Last Stand' came at the Great North American Eclipse of 29 July 1878, when the Moon's shadow cut a majestic swathe across the vast western United States, including through Yellowstone and the Wind River Range, down through Medicine Bow and south into Colorado, crossing Long's Peak, on through Boulder, Denver, Colorado Springs, Oklahoma Indian Country and finally leaving the United States and passing into the Gulf of Mexico between Galveston and New Orleans. Many astronomers were eager to use the opportunity to search for the intra-Mercurial planet, including the famed University of Michigan astronomer James Watson (1838–1880), who obtained the most sensational – if disputed – results.[9]

Watson had been born in Canada in 1838; he rose from adverse circumstances and, through self-directed study became, by the age of fifteen, a student at the University of Michigan in Ann Arbor. He soon gained a reputation as a skilful – and rapid – computer of comet and minor-planet orbits. He was known to be able to compute the elliptic elements of a minor planet's orbit at a single sitting. His amazing fluency in calculations allowed him for many years to supplement his income – and incur the envy of lesser-endowed astronomers – by moonlighting as an insurance actuary. He became a full professor of astronomy at the tender age of 21, and took charge of the university's observatory at 24. Within a matter of weeks he discovered a minor planet, Eurynome (no. 79). The euphoria he experienced proved addictive, and henceforth the discovery of minor planets became his passion. Sadly but perhaps inevitably he became a competitor, and eventually a bitter rival, of Christian Heinrich Friedrich Peters (1813–1890) of Hamilton College in rural New York. At the time Peters was himself the leading American discoverer of minor planets.

Both were flawed personalities. Watson was generous in his own circle, but did not seek out friends: 'For the ordinary forms of social intercourse he had no taste, and he held aloof from them, giving to his work hours that others spent in recreation.'[10] Peters was notoriously curmudgeonly, and possessed an unfortunate affinity for litigation. It is perhaps a comment on the bitterness of contemporary opinion that Peters's *Biographical Memoir* for the National Academy of Sciences, usually submitted shortly after a member's death by a close colleague and often expressing considerable affection, remained unwritten for over a century until it was finally removed from the backlog by the present author.[11]

A clash between the two was inevitable. When it came, it came over Vulcan. Watson had long believed in Vulcan. He had corresponded with Le Verrier, and at the latter's request had observed the Sun's disc at times when transits of the planet were predicted. Vulcan failed to show itself. Still, he remained hopeful, and thought that the forthcoming eclipse of the Sun – when the Sun would be hidden by the Moon for not quite three minutes – would offer a grand opportunity. With a 10-cm (4-in.) refractor borrowed from the Michigan State Normal School at Ypsilanti (now Eastern Michigan University), he planned to scan the stars of Cancer against which the Sun would be silhouetted. He knew this part of the sky well, having previously prepared a secretive set of charts of stars along the ecliptic to give him an advantage over his rivals in his minor planet searches. (Peters was, of course, in possession of his own secret set of charts.)

Watson took the train from Ann Arbor to Rawlins, Wyoming, a rough frontier town near the Continental Divide along the route pioneered by frontiersman Jim Bridger that would lie on the centreline of totality. For the first and only time in its existence, it was bustling with astronomers. J. Norman Lockyer (1836–1920) of England and pioneer spectroscopist and astro-photographer Henry Draper (1837–1882) of New York City were there.

(Mrs Draper, a New York City heiress, also came along, but would never see the eclipse; she was assigned by her husband and the other male astronomers to counting seconds out loud from within a tent, where she would not be distracted by the spectacular goings-on overhead!) Simon Newcomb (1835–1909), who had been recently appointed director of the United States Nautical Almanac office in Washington DC, was also there. Newcomb was a rival of Le Verrier in the calculation of tables of planetary motion, and was a veteran of the intra-Mercurial planet quest, having looked for them (in vain) at the 1869 eclipse. Even Thomas Edison, of Menlo Park, New Jersey, was there, toting one of his inventions, a pocket-sized device for measuring infrared radiation he called a tasimeter.

On the day of the eclipse, with winds high in Rawlins, Newcomb and Watson decided to try their luck at nearby Separation, an isolated Union Pacific rail stop at an elevation of 2,100 m (6,900 ft) some 24 km (15 mi.) from the summit of the Rocky Mountains.

Astronomical party at Rawlins, Wyoming, for the Great North American eclipse of 29 July 1878, when 'Vulcan fever' was at its highest pitch. James Watson is the stout figure sixth from the right; among others are, to his left, Mrs Watson, Mrs Henry Draper, Dr Henry Draper, Thomas Alva Edison (with arms folded), and J. Norman Lockyer.

Even in its prime, Separation consisted only of a wooden water tower, a rail siding and a few small wooden buildings where the station agent lived and worked, but at least there was some kind of accommodation for the astronomers. About 1.2 km (¾ mile) east of the station a semicircular sand dune, 4.6 m (15 ft) high, furnished protection from buffeting winds to the south and west. It was here that the planet-searchers set up their instruments.[12]

In advance of the eclipse, Watson had committed to memory the positions of all the stars to seventh magnitude, in a search zone centred on the Sun, 15 degrees long and 1½ degrees wide. In addition, because the telescope he had borrowed for the occasion was not ideally suited to the task demanded of it, he had to improvise. Though it was mounted equatorially, it was not furnished with setting circles, so he fitted the axes with circular covers of white cardboard, over which he mounted a pointer with a knife edge, intending to mark with a pencil the position of any suspicious object.

On eclipse day, and well before totality, Watson began sweeping east and west of the Sun, with an eyepiece magnifying 45x. As totality commenced, he placed the Sun in the middle of the field and moved the telescope slowly and uniformly to the east. He retraced his path, then moved the telescope one field to the south and began sweeping again. He encountered Delta Cancri and other known stars, re-centred on the Sun, and began sweeping in the same manner to the west. As he did so, between the Sun and Theta Cancri, he came across, as he recalled, 'a ruddy star whose magnitude I estimated to be 4½. It was fully a magnitude brighter than Theta Cancri, which I saw at the same time, and it did not exhibit any elongation, such as might be expected if it were a comet.'[13] (He later contradicted this statement, however, claiming that even at 45x it had a perceptible disc.) Watson used the pointer to mark the position on his circles, took down the chronometer time, and continued sweeping. Several degrees west of the Sun he encountered another, even brighter star, 'also ruddy in appearance', and marked its position, too, on the circles. Now seized with excitement, he ran over to Newcomb, 'in hopes that he might . . . get a place of the strange star which I had first observed'. Newcomb, however, was that moment absorbed in reading off his setting circles the position of a star he had in the field. It was north of the Sun, and would prove to be an ordinary star. Watson's two objects were south of the Sun, and though he conceded that one might be a known star on the chart, he felt sure that the other was a 'stranger'. (Newcomb could not help regretting afterwards that he did not let his own object go and point his telescope on Watson's.)

By the time Watson returned to his telescope, the Sun was already peeking out from behind the Moon's limb. The brief span of totality was over. Against the suddenly brightened sky, Watson was unable to recover his second object, and therefore could not decide whether it was a known star (Zeta Cancri) or not. Concerning the first ruddy star, however, he was confident; it was not a known star, and the record of the paper circles, carefully inspected by Newcomb and Lockyer in situ, would prove it.

There was nothing more to do but await the results of many other observers poised further along the eclipse track. In the end, there was to be only one seeming confirmation of Watson's planet, by Lewis Swift (1820–1913), a well-known discoverer of comets from Rochester, New York, who observed the eclipse from Denver. He offered a tantalizing report of two stars of approximately fifth magnitude, about eight minutes of arc apart. One was identified as Theta Cancri, the other was unknown, but Swift professed to having no doubt that it was an intra-Mercurian planet.

Unfortunately, closer scrutiny invited scepticism. Watson discovered several errors leading to corrections to the positions of the stars that he had observed. By late August, these revisions had led him to announce that the second object, which he had supposed to be Zeta Cancri, could not be that star after all but must also be an intra-Mercurian planet. Meanwhile, Swift published additional information about his observations, which were found to be incompatible with Watson's. Soon things were thrown into confusion; now, instead of one planet or even two, there had to be four – or none. Stock in Vulcan was plummeting.

Peters had been following developments closely. At last, he was ready to strike against his detested rival. He published a paper, 'Some Critical Remarks on So-called Intramercurial Planet Observations' in the journal *Astronomische Nachrichten*.[14] His main point was that Watson had overestimated the accuracy of his paper circle method of measuring the positions of his stars. As he saw it, Watson's lack of a telescope with proper setting circles had led to failure. 'The marking was done in the dim light of the total eclipse, or with lamplight,' Peters said. 'Either the slightest touch would bend the pointer, the flexible brass wire, a little to the side, or a parallax of some amount was unavoidable. The marking had to be done expeditiously and with a certain hurry.' The circumstances invited errors, and Peters concluded that Watson had observed Zeta and Theta Cancri, 'nothing else'. Peters's arguments prevailed, with the noted Irish historian of astronomy Agnes Clerke (1842–1907) issuing a verdict which has stood the test of time: 'The most feasible explanation of the puzzle seems to be that Watson and Swift merely saw each the same stars in Cancer: haste and excitement doing the rest.'[15]

The haste and excitement are real enough, and will be readily appreciated by anyone who has been present at a total eclipse of the Sun; the moments fly past, and as a leading historian of eclipse expeditions has said, 'the tension of eclipse observations in which delicate scientific equipment must work at a particular moment of time outside the observer's control is . . . unimaginable.'[16] Though occasionally someone muses that perhaps Watson's observation was too easily given up, and that he really did see something – a sun-grazing comet, for instance, or a coronal mass ejection – on the whole neither suggestion seems very likely.

Vulcan in Eclipse

The Vulcan myth did not die easily. Watson, who left the University of Michigan to become director of the observatory at the University of Wisconsin, Madison, spent the rest of his life obsessed with Vulcan. He even set up, at his own expense, an underground observatory from which he hoped to search for intra-Mercurial planets in broad daylight. It was still incomplete at the time of his death in 1880, of pneumonia contracted while installing a steam furnace in his house, and though completed by his successor, Edward Singleton Holden (1846–1914), it proved a complete failure. In 1883 Holden himself searched for Vulcan as head of an American party that went to tiny Caroline Island in the South Pacific. As usual, it failed to make an appearance, and he concluded that it would be futile to devote a telescope or observer to such purposes in the future. His conclusion: 'I must regard the fact of the non-existence of Vulcan as definitely settled.'[17]

Meanwhile, astronomical photography was advancing quickly, and by 1900 had completely supplanted the desperately hurried and unreliable searches of visual observers like Watson and Swift. At the 28 May 1900 solar eclipse William H. Pickering (1858–1938) of Harvard searched for Vulcan with a specially designed wide-angle lens, but without success. Teams from the Lick Observatory, using a specially designed 'Vulcan camera', carried out a series of photographic searches for the fugitive planet at the eclipses of 18 May 1901 (Sumatra), 30 August 1905 (Labrador, Spain and Egypt) and 3 January 1908 (Flint Island). On plates taken at the 1908 eclipse, three hundred stars in the neighbourhood of the Sun, down to a 9th magnitude, were recorded; all agreed in position with well-known stars. The Vulcan camera's reach would have turned up any bodies down to 50 km (30 mi.) in diameter. Since millions of such bodies were needed to furnish enough mass to explain the anomalous advance of Mercury's perihelion, the Lick director,

William Wallace Campbell (1862–1938), could confidently affirm, 'In my opinion, [this work] brings the observational side of the famous intra-Mercurial planet problem definitely to a close.'[18]

The Nagging Perihelion Problem and Its Resolution

Vulcan was dead; but the problem that it was invoked to explain lingered. In 1908, the anomalous advance of the perihelion of Mercury that had summoned the ghost seemed as far from solution as ever.

Simon Newcomb, having in 1882 repeated Le Verrier's Mercury calculations, confirmed that the anomalous advance of the perihelion was genuine, though he corrected the value slightly; instead of Le Verrier's 38″ per century he found 43″ per century, a result that would prove definitive.[19] A brief hope – or will-o'-the-wisp – of a reconciliation between negative observational results and theory was chased by United States Naval Observatory astronomer Asaph Hall (1829–1907), famed as the discoverer of the satellites of Mars. In 1894 Hall published a paper in which he tried tinkering with the inverse square law of gravitation itself. Instead of the exponent being exactly 2, he proposed it be amended to 2.00000016. Almost at once, however, it fell out that this adjustment led to unacceptable consequences in the motion of the Moon. Two years later, the German astronomer Hugo von Seeliger (1849–1924) invoked ellipsoidal concentrations of small particles around the Sun, possibly related to the tenuous 'Zodiacal Light'. Seeliger's attempt satisfied Newcomb that the Newtonian law of gravitation could be salvaged. However, in fact it was little more than an effort to sweep the problem under the carpet.

In the end, the astronomers would never solve the problem. The solution, when it arrived, came from a different quarter. The correct, non-Newtonian explanation was found by Albert Einstein (1879–1955) in November 1915 at a desk in Berlin. His work at the time was not aimed at solving the anomalous advance of Mercury's perihelion; rather he was working through the consequences of new gravitational equations derived from his General Theory of Relativity, and found that for a body like Mercury, lying close to a massive body like the Sun, the curvature of Einsteinian space-time introduces a significant non-Newtonian correction to the motion. The correction, he calculated, comes out to 0.1″ of arc for each orbital revolution or 43″ per century – identical to the value found by Newcomb. General Relativity had at last abolished the need for Vulcan – or indeed for any significant amount of circum-solar matter.

Albert Einstein lecturing in Vienna in 1921.

Indeed, the anomalous precession of the perihelion of Mercury predicted by General Relativity has now been confirmed to within the limits of observational error, and is significant enough to have to be taken into account for the accurate navigation of the MESSENGER spacecraft mission to Mercury. The subject of the confirmations of General Relativity is too vast to consider further here, but one deserves to be mentioned, as it applies specifically to Mercury. Jacques Laskar and Mickaël Gastineau at the Paris Observatory have carried out numerical solutions of the evolution of the Solar System with a supercomputer, and have found that there is some probability over the next few billion years of a change in the eccentricity of Mercury's orbit capable of producing collisions of Mercury with Venus and the Sun, or of Mercury, Mars or Venus with the Earth.[20] Without inclusion of the effects of General Relativity, none of these calculations would have been valid.

Though Vulcan does not exist, Le Verrier's other idea – that an intra-Mercurial band of asteroidal material might inhabit the zone in which its existence was once postulated – remains viable. It turns out that the region between 0.07 and 0.21 AU, the so-called 'vulcanoid zone', is gravitationally stable. Since other gravitationally stable zones in the Solar System contain objects such as planets and asteroids, remnants of a primordial vulcanoid population (planetesimals) might linger here, despite undoubted depletion by the combined effects of collisional erosion and radiative transport out of the zone. The zone might have been further stocked from the debris of a large impact on Mercury occurring relatively late in its evolution that at one point was postulated to have stripped away much of its crust and mantle and left its outsized iron core.

In addition to the intrinsic interest of this matter, the possibility of an intra-Mercurial asteroid belt led to serious concern over the validity of the cratering chronology of Mercury, over which planetary scientists toiled in the years after Mariner 10's success. Since the early days of geological mapping of the Moon by the United States Geological Survey in the early 1960s (first at Menlo Park, California, and later at Flagstaff, Arizona), one of the most important objectives was to work out the sequence of stratigraphic units laid down over the course of lunar geologic time. At first the schemes were very crude, and involved nothing more than division into 'pre-maria rocks', 'maria rocks' and 'post-maria rocks', but over time – especially after the adoption of the impact theory of crater origin – it was established that the abundance of craters on a given surface was a key to its relative age, and later, when the Apollo astronauts brought samples of Moon rocks back to Earth, the relative ages could be turned into absolute ages. In due course the calibration of lunar chronology was extended to other planets – notably Mars. Using Solar System dynamics and observations of Near-Earth Objects, the

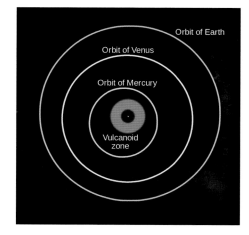

Vulcanoid Zone – a gravitationally stable region inside the orbit of Mercury, long postulated to contain a band of planetesimals (vulcanoids) left over from the origin of the Solar System.

Orbit of Earth

Orbit of Venus

Orbit of Mercury

Vulcanoid zone

lunar chronology could also be extended to Mercury, provided there was not a population of small bodies that would specifically strike Mercury but not the other planets, as 'vulcanoids' would do. Thus a careful inventorying of vulcanoids was a matter of considerable practical importance.

As we have seen, the surveys with the Lick Observatory's 'Vulcan camera' just after the turn of the twentieth century should have revealed any vulcanoids of more than 50 km (30 mi.) in diameter. A more recent study using NASA's Solar Terrestrial Relations Observatory Heliospheric Imager identified no vulcanoids down to an upper limit 5.7 km (3½ mi.) in diameter.[21] This upper limit was confirmed by MESSENGER, which imaged much of the region in which vulcanoids could have lasted dynamically from primordial times. Again, none was found, and the Mercury geologists concerned with working out Mercurian stratigraphy could breathe a collective sigh of relief. It now seems safe to extend the well-established lunar chronology to the geologic units of the innermost planet.

The Inside Story

Though the fact that Mercury is the innermost planet is one of those facts that everyone knows, and knows so well that it elicits no comment – every series, after all, must start with a first term – it innocently conceals one of the great mysteries of the Solar System. The situation in our Solar System, it turns out, is not quite normal. By analogy with the many exoplanet systems found since 1995, there ought to be a planet (or planets) closer in to the Sun than Mercury.

To understand wherein the enigma lies, consider the standard theory of the origin of the Sun and planets from the primordial solar nebula. During the Solar System's birth from a swirling cloud of dust and ice, the Sun formed as the most massive body at the centre. Close to the Sun, as the hot gases cooled, solid dust condensed, with the first condensates the refractory oxides, which can be studied directly in the carbonaceous meteorites that still preserve them. Further out, cooling progressed rapidly, and eventually reached the temperature at which water vapour condensed into ice. This is generally thought to have defined a 'frost line', at a distance of about 5 AU from the Sun – where Jupiter is now located and where, until recently, it was supposed it had first formed and thereafter remained. At still greater distances other more volatile molecules condensed – methane, ammonia and carbon dioxide – on grains of silicate dust, and at the very greatest distances, the most volatile ices formed nitrogen and carbon monoxide.

When described in these terms, it becomes obvious why the Solar System has the architecture it does: a few small rocky planets close to the Sun, several large gaseous planets further out, and the icy objects of the Kuiper Belt (including Pluto) at the icy outer rim.

Forming close to the Sun, embryonic Mercury ought to have been able to draw only materials with high-vaporization temperatures, such as iron. If it formed quickly, so the reasoning went, the increasing temperature of the Sun as it proceeded on its evolutionary track might have vaporized much of the existing crust of the planet, leaving only a thin shell. Alternatively, Mercury might have sustained an impact with another planet-sized body, and lost much of its original crust. These ideas seemed plausible until MESSENGER, whose surprising result was that Mercury contains much higher abundances of volatiles – sodium, potassium, chlorine and sulphur. High-temperature events such as heating by the Sun or a planetary impact would have driven the volatile elements off. Clearly, our theories of the formation of the inner planets need revision.

The first major revision is something that – though suspected before – has come into relief now that we know of other Solar Systems beyond our own. (And thanks to NASA's Kepler mission, thousands are now known.) When the first exoplanet systems were discovered, a number were found to have 'hot Jupiters' – giant planets unexpectedly close to their stars. This was entirely unexpected: a giant planet so close to the blistering heat of a star seemed like something out of Lewis Carroll – a 'bandersnatch' or 'borogrove' of a world, which somehow managed to beamishly 'gyre and gimble in the wabe' while growing to giant size despite the complete non-existence at such distances of giant-world-sustaining ice. These planets are way inside their solar systems' respective ice lines. Astronomers soon realized that the only possible explanation was that these giant planets had not formed so close to their stars, but further out – beyond the ice line – and had somehow migrated inwards.

The 'somehow' involves orbital resonances. We have already introduced the topic of resonances in some detail in the case of the almost 3:2 resonance between Mercury's spin and orbital motion. Another resonance, studied in detail by the eighteenth-century French doyen of celestial mechanics Pierre Simon de Laplace (1749–1827), involves a near 5:2 resonance between the mean motions of Jupiter and Saturn: Jupiter completes almost exactly five orbits in the time that Saturn completes two. Modern dynamical astronomers – in contrast to Laplace – appreciate the fact that giant planets, once formed, do not remain fixed in place, as the jewels in an eternal and unperishing celestial clockwork. Instead, they can drift around.[22]

In exoplanet systems with 'hot Jupiters', we see cases where the giant planets, though formed beyond the ice lines, have clearly drifted inwards towards their stars. As was first shown in 2001 by Frederic Masset and Mark Snellgrove, then at Queen Mary University, London, on the basis of computer simulations, when Jupiter and Saturn formed, they too were involved in inward migration across the Sun's protoplanetary disc, with Saturn, because of its lesser mass, migrating more rapidly.[23] Eventually, the two planets came closer together and reached a resonance – not the current near 5:2 resonance of today, but an exact 1:2 resonance. Owing to this resonance, Jupiter and Saturn exerted an amplified mutual gravitational influence on each other and on the remaining protoplanetary disc, producing a gap in the latter. Jupiter, with its greater mass, exerted more of a pull on the inner disc than Saturn on the outer. It may well have migrated inwards as far as the present orbit of Mars, scattering planetesimals before it and tossing them into the inner Solar System where they could seed the still unformed inner planets with water and other volatiles. The scattering of these planetesimals eventually caused Jupiter and Saturn to reverse their motion (the 'Great Tack') and migrate back out into the outer Solar System where they are today.

What is not clear from this scenario, first proposed on the basis of computer simulations by Kevin J. Walsh, then at the Côte d'Azur Observatory in Nice, France, and his colleagues as recently as 2011,[24] was why – in contrast to other exoplanet systems, many of which are packed with close-in 'super Earths' – the inner part of our Solar System appears hollowed out, why there are no planets closer in to the Sun than Mercury. A likely answer emerged in 2015, when Konstantin Batygin and Gregory Laughlin showed that

> Jupiter's inward migration . . . most likely triggered collisional cascade . . . that eroded the planetesimal population, essentially grinding them back down to boulders, pebbles and sand while pushing any primordial super-Earths into death spirals into the Sun. Computer simulations show that none of these hypothetical planets would have survived longer than a few hundreds of thousands of years. From the remaining relatively sparse and narrow ring of rocky debris, the terrestrial planets – Mercury, Venus, the Earth, and Mars – coalesced hundreds of millions of years later.[25]

Meanwhile, continuing to migrate outwards, Jupiter and Saturn encountered the recently formed Uranus and Neptune (and perhaps other gas giant planets that may later have been thrown clear of the Solar System altogether). The giant planets thus became locked into a chain of stabilizing resonances where they could remain for billions of years – perhaps resembling the four giant planets in the recently discovered Kepler-223 system, which occupy such a chain of resonances in which for every three orbits of the outermost one, the second orbits four times, the third six times and the innermost eight times.[26]

In our Solar System things did not end there. The cumulative effect of gravitational interactions between the outer planetesimals and the giant planets eventually shifted the latter, causing Jupiter to edge slightly inwards and the rest of the giant planets outwards. The result was that the chain of resonances that had previously stabilized them broke down, and the giant planets were thrown from a well-ordered series of closely spaced circular orbits into a chaotic jumble of widely spaced and wildly eccentric ones. (It is at this stage that one – or more – giant planets may have been ejected from the Solar System and flung into interstellar space.) As the giant planets – and especially Jupiter – followed their eccentric paths, they gave up some of the energy of their motion to any planetesimals that came close. Most of these were ejected out of the Solar System, but some were tossed through the inner Solar System to give rise to the so-called Late Heavy Bombardment that occurred between 4.1 billion and 3.8 billion years ago – an era that is well attested in the battered surfaces of the inner planets. Mercury's Caloris basin and other basins are scars from this violent era of interplanetary ballistics. Eventually the energy exchange between the motions of the giant planets and the planetesimals recircularized their orbits, at the end of this period leaving the Solar System effectively as it is today.

Before the turbulent period ended, a number of inner planets might well have been tossed into the Sun. Mercury was the innermost world that survived. Since then it has continued to endure the scorching heat of the Sun and the intense blasts of the solar wind, and in spite of all continues to hail us – during the brief interludes of pre-sunrise or post-sunset twilight – as a twinkling sentinel on the Solar System's inner frontier.

APPENDIX 1:
GLOSSARY

Albedo – The fraction of incident light reflected by a surface.

Altazimuth – Moving parallel to the horizon and perpendicular to the horizon.

Arc second – An angular unit equal to 1/3600 of one degree.

Astronomical seeing – the blurring and twinkling of astronomical objects due to turbulence in the Earth's atmosphere.

Cassegrain – A design of reflecting telescope in which light collected by the primary mirror is reflected to a small secondary mirror on the optical axis, which, in turn, reflects the light through a hole in the primary. This allow the eye piece to be placed behind the telescope.

CCD – Charged Couple Device. An array of light-sensitive electronic chips used to produce an image.

Celestial Meridian – An imaginary semicircle running North–South through the zenith. It divides the sky into an eastern half and a western half.

Conjunction – Two astronomical objects appearing in the same location (or nearly the same location) on the Celestial Sphere.

Deferent – In the Ptolemaic theory of the planetary motions, a larger circle round which an epicycle moves.

Distal – The outer zones of an area affected by geologic activity.

Doppler shift – The wavelength change that occurs when a wave-emitting object moves radially relative to the observer. Perceived wavelengths appear shorter as observer and emitter approach each other; they appear longer as observer and emitter recede from one another.

Eccentricity – A measure of how much an elliptical shape deviates from that of a circle. It is a unit-less number equal to the distance between the two foci of the ellipse divided by the length of the major axis. The eccentricity of a circle is equal to 0, that of a very stretched-out ellipse approaches 1.

Ecliptic – The plane of the Earth's orbit around the Sun. It is used as a reference from which the inclination of other bodies in solar orbit is measured.

Elongation – From the Earth, the angular distance between an astronomical object and the Sun.

Epicycle – A small circle round which a planet moves, the centre of which is itself carried around a larger circle, 'the deferent'.

Graben – A depressed area of land bordered by two vertical faults.

Inferior conjunction – The passage of an inferior planet, such as Mercury, between the Earth and the Sun.

Inferior planet – Mercury or Venus, so-called because they are closer to the Sun than is the Earth.

Lobate – Having a lobe or lobes.

Newtonian – A design of reflecting telescope in which light collected by the primary mirror is reflected to a small secondary mirror on the optical axis, which, in turn, reflects the light to the side of the telescope, where an eye piece is placed.

Precession – The slow wobble of a planet's rotational axis due to the perturbing gravitational force of other bodies. The period of precession for the Earth is 26,000 years.

Ray – A feature of impact craters that occurs when higher-albedo material beneath the dark surface of a solar-system body is scattered radially away from the impact site.

Reflector – A reflecting telescope. A concave mirror is used to collect light.

Refractor – A refracting telescope. A lens is used to collect light.

Retrograde – The usual, prograde motion of solar-system bodies on the Celestial Sphere is west to east. When a body appears to be moving east to west, its motion is said to be retrograde.

Setting circles – Every location on the Celestial Sphere can be specified by two coordinates, called right ascension and declination. When a telescope is mounted such that one of its axes of motion is parallel to the axis of the Earth, and the other perpendicular to it, circular scales may be placed on the telescope axes that correspond to right ascension and declination. By setting these circles to a known right ascension and declination, the telescope will point to that location in the sky.

Solar wind – The continuous emission of charged electrical particles from the Sun.

Superior conjunction – An inferior planet such as Mercury is located on the opposite side of the Sun from us.

Synodic period – The time it takes for a solar-system body to make one revolution in the sky, with respect to the Sun.

Transit – The passage of a smaller body, such as Mercury, across the disc of a larger one.

Zenith – The point directly overhead.

APPENDIX II:
BASIC DATA

Discovery

Date of Discovery: Unknown

Discovered By: Known by the Ancients

Orbit Size around Sun

Metric: 57,909,227 km

English: 35,983,125 miles

Scientific Notation: 5.7909227×10^7 km

Astronomical Units: 0.38709927 AU

By Comparison: Earth is 1 AU (Astronomical Unit) from the sun.

Mean Orbit Velocity

Metric: 170,503 km/h

English: 105,946 mph

Scientific Notation: 4.7362×10^4 m/s

By Comparison: $1.590 \times$ Earth

Orbit Eccentricity

0.20563593

By Comparison: $12.305 \times$ Earth

Equatorial Inclination

0 degrees

By Comparison: Earth's equatorial inclination to orbit is 23.45 degrees.

Equatorial Radius

Metric: 2,439.7 km

English: 1,516.0 miles

Scientific Notation: 2.4397×10^3 km

By Comparison: $0.3829 \times$ Earth

Equatorial Circumference

Metric: 15,329.1 km

English: 9,525.1 miles

Scientific Notation: 1.53291×10^4 km

By Comparison: $0.3829 \times$ Earth

Volume

Metric: 60,827,208,742 km³

English: 14,593,223,446 miles³

Scientific Notation: 6.08272×10^{10} km³

By Comparison: $0.056 \times$ Earth

Mass

Metric: 330,104,000,000,000,000,000,000 kg

Scientific Notation: 3.3010×10^{23} kg

By Comparison: $0.055 \times$ Earth

Density

Metric: 5.427 g/cm³

By Comparison: $0.984 \times$ Earth

Surface Area
Metric: 74,797,000 km^2
English: 28,879,000 square miles
Scientific Notation: 7.4797×10^7 km^2
By Comparison: $0.147 \times$ Earth

Surface Gravity
Metric: 3.7 m/s^2
English: 12.1 ft./s^2
By Comparison: If you weigh 100 lb on Earth, you would weigh 38 lb on Mercury.

Escape Velocity
Metric: 15,300 km/h
English: 9,507 mph
Scientific Notation: 4.25×10^3 m/s
By Comparison: Escape Velocity of Earth is 25,030 mph

Sidereal Rotation Period
58.646 Earth Days
1407.5 Hours
By Comparison: $58.81 \times$ Earth

Surface Temperature
Metric: -173/427 °C
English: -279/801 °F
Kelvin Scale: 100/700 K
By Comparison: Earth's temperature range is ~ 185/331 K.

Atmospheric Constituents
By Comparison: Earth's atmosphere consists mostly of N$_2$ and O$_2$.

APPENDIX III:

CRATERS

Craters	Diameter (km)	Centre Latitude	Centre Longitude	Origin of Name
Aksakov	174	34.71	78.74	Sergey; Russian author (1791–1859)
Al-Hamadhani	164	39.19	91.76	Arab writer (d.1007)
Alver	151.49	-66.97	282.75	Betti; Estonian poet (1906–1989)
Aneirin	467	-27.47	2.68	Welsh poet (fl6th century)
Bach	214.29	-69.86	103.01	Johann Sebastian; German composer (1685–1750)
Beethoven	630	-20.86	124.21	Ludwig van; German composer of Flemish descent (1770–1827)
Bernini	168.13	-80.32	140.97	Gian Lorenzo; Italian sculptor and architect (1598–1680)
Boccaccio	151.95	-80.99	23.01	Giovanni; Italian poet and novelist(1313–1375)
Bramante	156	-47.23	61.55	Donato; Italian architect (1444–1514)
Caravaggio	185	3.18	272.76	Michelangelo Merisi; Italian painter (1571–1610)
Catullus	1002	21.88	67.56	Gaius Valerius; Roman poet (c.84–c.54 BCE)
Cervantes	213.16	-76.09	124.26	Miguel de; Spanish novelist, playwright and poet (1547–1616)
Chaikovskij	171	7.86	50.93	Pyotr Ilyich; Russian composer (1840–1893); also sometimes Tchaikovsky
Chao Meng-Fu	140.73	-88.42	156.36	Chinese painter and calligrapher (1254–1322)
Chekhov	194	-36.2	61.23	Anton; Russian playwright and writer (1860–1904)
Chŏng Ch'ŏl	143	46.87	117.31	Korean poet (1536–1593)
Copland	208	37.63	287.01	Aaron; American composer and pianist (1900–1990)
Dalí	176	45.16	240.26	Salvador; Spanish painter (1904–1989)
Darío	151	-26.28	9.51	Ruben; Nicaraguan poet, journalist (1867–1916)
Delacroix	158	-44.32	129.51	Eugène; French painter (1798–1863)

Craters	Diameter (km)	Centre Latitude	Centre Longitude	Origin of Name
Derain	167	-9	340.3	André; French painter (1880–1954)
Derzhavin	156	45.6	36.93	Gavrila Romanovich; Russian poet (1743–1816)
Dostoevskij	430.06	-44.73	178.11	Fyodor Mikhaylovich; Russian novelist (1821–1881); also sometimes Dostoevsky
Dowland	158	-53.56	180.7	John; English composer (1562–1626)
Dürer	195	21.51	118.98	Albrecht; German painter (1471–1528)
Ellington	216	-12.88	333.9	Edward Kennedy 'Duke'; American composer, musician, conductor (1899–1974)
Faulkner	168	8.08	283.03	William; American author (1897–1962)
Giotto	144	12.46	56.48	Italian painter (c. 1271–1337)
Goethe	317.17	81.1	51.03	Johann Wolfgang von; German poet and dramatist (1749–1832)
Hafiz	280	19.5	280.44	Hafiz (or Hafez) of Shiraz, Shams-ud-din Muhammad; Persian poet (c. 1320–1389)
Haydn	251	-27.22	71.64	Joseph; Austrian composer (1732–1809)
Henri	163.8	79.68	207.02	Robert; American painter (1865–1929)
Holst	170	-17.42	315.04	Gustav Theodore; British composer (1874–1934)
Homer	319	-1.3	36.62	Greek epic poet (8th or 9th century BCE)
Hugo	206	39.61	48.49	Victor; French writer, dramatist and poet (1802–1885)
Ibsen	159	-24.35	35.89	Heinrich J.; Norwegian poet and dramatist (1828–1906)
Imhotep	159	-17.97	37.48	Egyptian physician and sage (c. 2686–2613 BCE)
Izquierdo	174	-1.66	252.96	María; Mexican painter (c. 1902–1955)
Jobim	167	32.45	66.88	Antonio Carlos; Brazilian composer and musician (1927–1994)
Kālidāsā	160	-18.21	179.82	Indian poet and dramatist (c. 5th century)
Kipling	164	-19.37	287.98	Rudyard; English author (1865–1936)

Craters	Diameter (km)	Centre Latitude	Centre Longitude	Origin of Name
Kuan Han-Ch'ing	143	29.44	53.67	Chinese dramatist (c. 1241–1320)
Kunisada	241.45	1.37	246.98	Utagawa Kunisada; Japanese woodblock printmaker (1786–1864)
Kurosawa	152	-52.44	21.49	Kinko; Japanese composer (18th century)
Lange	176.23	6.28	259.53	Dorothea; American photographer (1895–1965)
Larrocha	196	43.29	69.83	Alicia de Larrocha; Spanish pianist (1923–2009)
Lermontov	166	15.24	48.94	Mikhail Yurevich; Russian poet (1814–1841)
Liang K'ai	145	-39.89	184.16	Chinese painter (c. 1140–1210)
Lysippus	155	1.03	132.75	Greek sculptor (4th century BCE)
Ma Chih-Yuan	197	-60.01	78.01	Chinese dramatist (fl. 1251)
Magritte	149	-72.78	238.37	René; Belgian painter (1898–1967)
Mark Twain	142	-10.91	138.28	(Samuel Clemens); American novelist, satirist (1835–1910)
Matisse	189.04	-23.83	90.05	Henri; French painter and sculptor (1869–1954)
Melville	146	22.01	9.89	Herman; American novelist (1819–1891)
Mendelssohn	291	70.31	257.68	Jakob Ludwig Felix; German composer (1809–1847)
Mendes Pinto	192	-61.65	17.57	Fernão; Portuguese prose author (c. 1510–1583)
Michelangelo	229.71	-44.92	109.78	Buonarroti; Italian painter, sculptor and architect (1475–1564)
Milton	180.85	-26.1	175.03	John; English poet (1608–1674)
Monet	203	44.23	9.77	Claude; French painter (1840–1926)
Mozart	241	7.75	190.59	Wolfgang Amadeus; Austrian composer (1756–1791)
Munkácsy	193	21.95	258.86	Mihály; Hungarian painter (1844–1900)
Nabokov	166	-14.56	304.24	Vladimir; Russian and American writer (1899–1977)
Petrarch	167	-30.52	26.29	Francesco; Italian poet (1304–1374)
Phidias	168	8.97	149.73	Greek sculptor (fl. c. 490–430 BCE)
Pigalle	153	-37.65	9.64	Jean-Baptiste; French sculptor (1714–1785)

Craters	Diameter (km)	Centre Latitude	Centre Longitude	Origin of Name
Praxiteles	198	27.11	60.28	Greek sculptor (fl. 370–330 BCE)
Proust	145	19.56	47.59	Marcel; French novelist (1871–1922)
Pushkin	232	-65.79	20.73	Alecksandr Sergeyevich; Russian poet (1799–1837)
Rabelais	154	-60.53	61.81	François; French writer (c. 1483–1553)
Rachmaninoff	305	27.66	302.63	Sergei V; Russian composer, pianist and conductor (1873–1943)
Raditladi	258	27.15	240.94	Leetile Disang; Botswanan playwright and poet (1910–1971)
Raphael	342	-20.42	76.35	Raffaello Sanzio; Italian painter (1483–1520)
Rembrandt	716	-32.89	272.13	Rembrandt Harmenszoon van Rijn; Dutch painter (1606–1669)
Renoir	220	-18.34	52.01	Pierre Auguste; French painter (1841–1919)
Riemenschneider	183	-52.75	99.95	Tilman; German sculptor (c. 1460–1531)
Rodin	230	21.72	18.89	Auguste; French sculptor (1840–1917)
Rubens	158	60.58	77.99	Peter Paul; Flemish painter (1577–1640)
Rustaveli	200	52.41	277.26	Shota; Georgian poet (c. 1160–1216)
Sanai	490	-13.37	6.99	Sanai of Ghazna, Abul Majd bin Majdud bin Adam; Persian poet (d. c. 1131)
Sayat-Nova	146	-27.98	122.7	Aruthin Sayadian; Armenian/Georgian song writer (1712–1795)
Schubert	190	-43.21	54.26	Franz P.; Austrian composer (1797–1828)
Shakespeare	399	48.1	152.25	William; English poet and dramatist (1564–1616)
Shelley	171	-47.69	128.27	Percy Bysshe; English poet (1792–1822)
Shevchenko	143	-53.64	46.02	Taras Hryhorovych; Ukrainian poet (1814–1861)
Sholem Aleichem	196	50.92	90.48	(Yakov Rabinowitz); Yiddish writer (1859–1916)
Smetana	191	-48.25	70.17	Bedrich; Czechoslovakian composer (1824–1884)
Sophocles	142	-6.95	146.04	Greek dramatist (c. 496–406 BCE)
Sōtatsu	157	-48.73	18.17	Tawaraya; Japanese artist (1600–1643)

Craters	Diameter (km)	Centre Latitude	Centre Longitude	Origin of Name
Steichen	196	-12.79	282.96	Edward; American photographer and painter (1879–1973)
Strindberg	189	53.41	136.67	August; Swedish playwright, novelist and short-story writer (1849–1912)
Sullivan	153.23	-16.19	86.88	Louis; American architect (1856–1924)
Surikov	224	-36.98	124.9	Vassily; Russian painter (1848–1916)
Sveinsdóttir	212.79	-2.83	259.68	Júlíana; Icelandic painter and textile artist (1889–1966)
Tolstoj	355	-16.23	164.64	Leo Tolstoy; Russian novelist (1828–1910)
Vālmiki	210	-23.58	141.41	Sanskrit poet, author of the *Ramayala* (1st century BC)
Van Eyck	271	43.22	159.43	Jan; Flemish painter (c. 1395–1441)
Verdi	145	64.36	169.71	Giuseppe; Italian composer (1813–1901)
Vieira da Silva	274	1.54	123.41	Maria Helena; Portuguese-born French painter (1908–1992)
Vivaldi	213	13.76	85.92	Antonio; Italian composer (1678–1741)
Vyāsa	297	49.79	84.62	Indian poet (fl. 1500 BCE)
Wang Meng	165	8.53	104.12	Chinese painter (1308–1385)
Wren	204	24.84	35.95	Christopher; English architect (1632–1723)

References

1 The Scintillating One

1 E.-M. Antoniadi, The Planet Mercury, trans. Patrick Moore (Shaldon, 1974), p. 9.

2 Motions of Mercury

1 Owen Gingerich, The Eye of Heaven (New York, 1993), p. 55.
2 Richard Baum and William Sheehan, In Search of Planet Vulcan: The Ghost in Newton's Clockwork Universe (New York, 1997), p. 11.
3 On transits of Mercury and Venus, see John Westfall and William Sheehan, Celestial Shadows: Eclipses, Transits, and Occultations (New York, 2015), especially Chapters Nine and Ten.
4 Jean Meeus, Transits (Richmond, VA, 1989).
5 Camille Flammarion, Popular Astronomy: A General Description of the Heavens, trans. John Ellard Gore (London, 1894), p. 359.

3 Through the Telescope

1 Bernard le Bovier de Fontenelle, Conversations on the Plurality of Worlds, trans. H. A. Hargreaves (Berkeley, CA, 1990), p. 49.
2 The definitive source on the subject is John Westfall and William Sheehan, Celestial Shadows: Eclipses, Transits, and Occultations (New York, 2015).
3 E.-M. Antoniadi, The Planet Mercury, trans. Patrick Moore (Shaldon, 1974), p. 20.
4 On Schröter's lunar work, see William P. Sheehan and Thomas A. Dobbins, Epic Moon: A History of Lunar Exploration in the Age of the Telescope (Richmond, VA, 2001), especially Chapter Six.
5 Henry C. King, The History of the Telescope (New York, 1979, reprint of 1955 edition), p. 135.

6 Schröter published his observations of Mercury, together with others of the asteroid Vesta, in the final year of his life. The book is Johann Schröter, *Hermographische Fragmente zur genauren Kenntniss des Planeten Mercur. Mit Beobachtungen öber den neu entdeckten Planeten Vesta* (Gottingen, 1816).

7 Quoted in Richard Baum, 'The Lilienthal Tragedy', *Journal of the British Astronomical Association*, CI (1991), p. 369.

8 John Browning, 'The Condition of Jupiter', *The Student and Intellectual Observer* (February 1870), p. 5.

9 Richard Baum, 'An Observation of Mercury and its History', *Journal of the British Astronomical Association*, CVII (1997), p. 38.

10 Karl Zöllner, *Photometrische Untersuchungen uber die physische Beschaffenheit des Planeten Merkur* (Leipzig, 1874).

11 According to Richard A. Proctor (completed by A. Cowper Ranyard), *Old and New Astronomy* (London, 1892), p. 427.

4 ROTATION

1 For a biographical essay of Denning that breaks new ground, and to which we are much indebted here, see Martin Beech, 'W. F. Denning – The Doyen of Amateur Astronomers', *WGN, the Journal of the International Meteor Organisation*, XXVI/I (1998), p. 19.

2 H. G. Wells, *The War of the Worlds* (New York, 2005), p. 13. Other astronomers who are mentioned by name by Wells are Giovanni Schiaparelli and Henri Perrotin of the Nice Observatory, both not only keen observers of Mars but also of Mercury.

3 William F. Denning, *Telescopic Work for Starlight Evenings* (London, 1891), p. 137.

4 Phrases first used by Percival Lowell in one of his observing logbooks.

5 David L. Graham, 'The Nature of Albedo Features on Mercury, with Maps for the Telescopic Observer. Part II: The Nature of the Albedo Markings', *Journal of the British Astronomical Association*, CV/2 (1995), p. 59. Graham notes: 'Situated at longitude 127°, latitude 37°, this 45 km crater radiates a pattern of bright streaks up to 900 km across Sobkou Planitia with a 200-km long bright swath on its eastern flank, and is rivaled in size only by the crater Kuiper.'

6 Denning, *Telescopic Work*, p. 140.

7 Quoted in E.-M. Antoniadi, *The Planet Mercury*, trans. Patrick Moore (Shaldon, 1974), p. 26.

8 Ibid., p. 27.

9 As Denning noted in the chapter on the inferior planets, which he contributed to T.E.R. Phillips and W. H. Steavenson (eds), *Splendour of the Heavens* (London, 1923), p. 192.

10 A brother, Celestine, was a scholar of Arabic, and collected books and manuscripts in Persian and Sanskrit as well as Arabic. He also regularly exchanged coins from his collection with Umberto I, the second king of Italy and an enthusiastic numismatist. His daughter, Elsa Schiaparelli, became a famous fashion designer. One of their sisters was placed in charge of all the convents in Italy, while a cousin of Giovanni and Celestine was a noted Egyptologist who discovered Queen Nefertari's tomb in the Valley of the Queens.

11 Quoted in Piero Bianucci, 'Giovanni Virginio Schiaparelli', L'Astronomia, VI (1980), p. 45.

12 As a student in Turin, Schiaparelli made detailed notes of the lectures in his extremely small, precise and meticulous hand, and later copied them into what are essentially complete treatises based on these lectures (e.g. 'Lezioni sopra la curve . . .'). These were not the rough notes of a haphazard student but carefully rewritten and organized for future reference. He seems to have got on well with most of his professors; the one exception was the eminent but misanthropic Giovanni Antonio Amedeo Plana, famed for his researches on the mathematical theory of the Moon, whom Schiaparelli approached hoping to learn something of celestial mechanics. Plana did not have time for a student, and dismissed him brusquely with the words: 'We need only one astronomer here.'

13 A younger brother, Eugenio Schiaparelli, was killed fighting the Austrians at the battle of Solferino.

14 Percival Lowell, Mars and Its Canals (New York, 1906), p. 11. It is interesting to note – for those who see meaning in such coincidences – that they were born only 80 km (50 miles) apart, Columbus in Genoa, Schiaparelli in Savigliano.

15 G. V. Schiaparelli, 'Sulla di Mercurio', Astronomische Nachrichten, CXXIII/2944 (1890), p. 241. The English translations from this work are by William Sheehan.

16 His niece, the future leader of haute couture, Elsa Schiaparelli, claims in her autobiography Shocking Life (New York, 1954), p. 25: 'After his marriage he took his young bride to Vienna. On the evening of their arrival he exclaimed, "Excuse me, I must go and see an astronomer who lives in this town. I won't be long." He rushed off with his mind already full of telescopes and stars while his poor little wife stayed sobbing in the hotel bedroom. She waited for dinner. He did not come back. The hour for supper passed. Midnight struck. And then the whole night went by without any sign of Giovanni. The next morning my uncle, quite unconcerned, returned to the hotel, asked for the key, and went up. There, on the floor, a pathetic bundle, was his sweet wife still sobbing. The astronomer let out a cry of surprise. "Oh!" he exclaimed, running at her. "I completely forgot I was married!"'

17 Schiaparelli, Shocking Life, p. 26.

18 Ibid.

19 He wrote to his friend François Terby in 1879 that his eyes 'have only a little sensitivity to the nuances of colour'. G. V. Schiaparelli, *Corrispondenza su Marte*, I (Pisa, 1963), p. 28.

20 G. V. Schiaparelli, 'Sulla Rotazione e sulla costituzione fisica del pianeta Mercurio', *Atti della Reale Accademia dei Lincei* (1889), in G. V. Schiaparelli, *Le Opere di G.V. Schiaparelli*, 10 vols (Milano, 1930), vol. 5, p. 344.

21 The maximum elongation on 6 February was 18°.2 E of the Sun.

22 Note from observing logbooks; Brera Observatory.

23 Schiaparelli, 'Sulla di Mercurio', p. 241.

24 On psychological issues in planetary observation, see William Sheehan, *Planets and Perception: Telescopic Views and Interpretations, 1609–1909* (Tucson, AZ, 1988), and G. Berlucchi, 'Some Historical Crossroads between Astronomy and Visual Neuroscience', *Memorie della Società Astronomica Italiana*, LXXXII (2011), p. 225.

25 Thus, on 27 January 1895, he wrote to his friend Terby that during the 1890s he had been able to observe very little on Mars, being especially troubled by 'a diminution of the sensibility to weak illuminations; I attribute this to the observations of Mercury near the Sun.' Giovanni Schiaparelli, *Corrispondenza su Marte*, II (Pisa, 1976), p. 213.

26 Schiaparelli, 'Sulla Rotazione e sulla costituzione fisica del pianeta Mercurio', p. 247.

27 Terby to Schiaparelli, 12 April 1889, *Corrispondenza su Marte*, I (1963).

28 Translation from the Latin by William Sheehan.

29 Schiaparelli to Terby, 6 November 1883, *Corrispondenza su Marte*, I, p. 105.

30 Schiaparelli, 'Sulla Rotazione', p. 249.

31 Ibid., p. 337.

32 G. V. Schiaparelli, 'The Rotation and Physical Condition of the Planet Mercury', in *Essays in Astronomy by Ball, Harkness, Herschel, Huggins, Laplace, Mitchel, Proctor, Schiaparelli and Others, with a Critical Introduction by Edward Singleton Holden*, trans. Sara Carr Upton (New York, 1900), p. 133.

33 Ibid., p. 140.

34 Antoniadi, *The Planet Mercury*, p. 28.

35 Rudaux later became celebrated for his famous paintings of space-related themes, mostly depicting imaginary views from the lunar surface, in the 1920s and 1930s, and is remembered today as the most important pioneer of 'Space Art' before Chesley Bonestell.

36 By far the best biographical account of Antoniadi is Richard McKim, 'The Life and Times of E. M. Antoniadi, 1870–1944, Part I: An Astronomer in the Making', *Journal of the British Astronomical Association*, CIII/4 (1993), p. 164 and 'Part II: The Meudon Years', p. 219.

37 See William Sheehan, *The Planet Mars: A History of Observation and Discovery* (Tucson, AZ, 1996) and *Planets and Perception.*

38 E.-M. Antoniadi, 'Le Retour de la planète Mars', *L'Astronomie*, XL (1926), p. 345.

39 Antoniadi, *The Planet Mercury*, p. 28.

40 Edison Pettit and Seth B. Nicholson, 'Measurements of the Radiation from the Planet Mercury', *Publications of the Astronomical Society of the Pacific*, XXXV/206 (1923), p. 194.

41 E.-M. Antoniadi, 'The Markings and Rotation of Mercury', *Journal of the Royal Astronomical Society of Canada*, XXVII (1933), p. 402.

42 Ibid.

43 Antoniadi, *The Planet Mercury*, p. 30.

44 Ibid., p. 37. The difference in the appearance of the Mercurian markings no doubt owed much to the greater power of Antoniadi's telescope. Schiaparelli himself had fully recognized that what he saw as 'crisscrossing bands' and spots did not represent the true character of the Mercurian markings. Thus, in his observing logbook, he entered the following notes on 25 April 1883, when the figure-of-5 markings were present to view (quoted in Schiaparelli, 'Sulla Rotazione', p. 247): 'Unless I am mistaken, the figures here represented are in part subjective; they depend upon the way in which the eye tries to group complicated lines and small spots, which is what the streaks and larger spots would become if only one had greater optical power.'

45 Antoniadi, *The Planet Mercury*, p. 31.

46 Ibid.

47 E.-M. Antoniadi, 'The Atmosphere of Mercury and its Clouds', *Journal of the British Astronomical Association*, XLV/6 (April 1935), p. 236.

48 Antoniadi, *The Planet Mercury*, p. 63.

49 Audouin Dollfus, 'Visual and Photographic Studies of the Planets at the Pic du Midi', *Planets and Satellites*, ed. Gerard P. Kuiper and Barbara M. Middlehurst (Chicago, IL, 1961) p. 550.

50 Clark R. Chapman, *Planets of Rock and Ice: From Mercury to the Moons of Saturn* (New York, 1982), p. 68.

51 Ibid., p. 69.

52 G. H. Pettingill and R. B. Dyce, 'A Radar Determination of the Rotation of the Planet Mercury', *Nature*, CCVI (19 June 1965), p. 1240.

53 Personal communication with the author, 6 January 2017.

54 Dale P. Cruikshank and Clark R. Chapman, 'Mercury's Rotation and Visual Observations', *Sky & Telescope* (July 1967), p. 24.

55 Personal communication with the author and Thomas Hockey, 1 September 2016.

56 See W. Sheehan, J. Boudreau and A. Manara, 'A Figure in the Carpet: Giovanni Schiaparelli's Classic Observations of Mercury Reconsidered in

the Light of Modern CCD Images', *Memorie della Società Astronomica Italiana*, LXXXII (2011), p. 358, and W. Sheehan, J. Boudreau and A. Manara, 'The Mercury Mirage: One of Giovanni Schiaparelli's Most Celebrated Telescopic Discoveries is Reconsidered in the Light of Modern CCD Images', *Sky & Telescope* (March 2011), p. 30.

57 Sheehan, *Planets and Perception*, p. 85.

5 MERCURY UP CLOSE

1 Bruce Murray, *Journey into Space: The First Thirty Years of Space Exploration* (New York and London, 1989), p. 98.

2 Ibid., p. 105.

3 Ibid., p. 116.

4 Murchie et al., 'Geology of the Caloris Basin, Mercury: A View from MESSENGER', *Science*, CCCXXI (4 July 2008), p. 73.

5 Richard Baum, 'The Craters of Mercury: Possible Telescopic Sightings', *Journal of the British Astronomical Association*, XCVI (1986), p. 78.

6 See Ulrich R. Christensen, 'A Deep Dynamo Generating Mercury's Magnetic Field', *Nature*, CDXLIV (21 December 2006), p. 1056.

7 This is especially effective on Mercury, which is close to the Sun and is bombarded by interplanetary dust particles at extremely high velocities.

8 Clark Chapman, personal communication to W. S., 1 September 2016.

9 Dale Cruikshank, personal correspondence to W. S., 10 February 2008.

10 Personal communication from Dale Cruikshank.

11 Personal communication from David Graham.

12 See Brett W. Denevi, Mark S. Robinson, Sean C. Solomon et al., 'The Evolution of Mercury's Crust: A Global Perspective from MESSENGER', *Science*, CCCXXIV (1 May 2009), p. 613.

13 By comparison, most C-type asteroids are darker than Mercury; most of them much darker, with albedos of 2–5 per cent. Some comets are also very dark.

14 William E. McClintock, Ronald J. Vervack Jr., Todd Bradley et al., 'MESSENGER Observations of Mercury's Exosphere: Detection of Magnesium and Distribution of Constituents', *Science*, CCCXXIV (1 May 2009), p. 610.

6 VULCAN

1 The definitive biography on Le Verrier is James Lequeux, *Le Verrier – Savant magnifique et détesté* (Paris, 2009), which is also available in English as *Le Verrier – Magnificent and Detestable Astronomer*, trans. Bernard Sheehan, ed. with introduction by William Sheehan (New York, 2013).

2 U.-J.-J. Le Verrier, 'Détermination nouvelle de l'orbite de Mercure et de ses perturbations', *Comptes Rendus de l'Academie des Sciences*, XVI (1843), p. 1054.

3 U.-J.-J. Le Verrier, *Théorie du Mouvement de Mercure* (Paris, 1845).

4 O. M. Mitchel, *Orbs of Heaven* (London, 1857), p. 138.

5 Norwood Russell Hanson, 'Leverrier: The Zenith and Nadir of Newtonian Mechanics', *Isis*, LIII (1962), p. 359.

6 Urbain-Jean-Joseph Le Verrier, 'Théorie du Mouvement de Mercure', *Annales de l'Observatoire Impérial de Paris (Mémoires)*, V (1879), p. 105.

7 On Lescarbault, see Richard Baum and William Sheehan, *In Search of Planet Vulcan: The Ghost in Newton's Clockwork Universe* (New York, 1997), pp. 145–61, and Richard Baum, 'The Lescarbault Legacy', *Patrick Moore's Yearbook of Astronomy 2013* (London, 2012), p. 179.

8 See Lequeux, *Le Verrier*, pp. 167–8. Out of curiosity, I investigated the possibility that the Sun might have been photographed on this date. After all, the first daguerreotype image of the Sun was obtained by Leon Foucault (1819–1868) and Louis Fizeau (1819–1896) as early as 1845, while the solar photography programme at Kew Observatory, near London, began in 1858, though owing to funding shortages it was not until 1862 that the photoheliograph was used on a daily basis. Thus, it seemed there was at least a possibility that a photograph might have been taken on 26 March 1859. At my request, Dr Lee MacDonald kindly made a search, but though he finds that the Royal Greenwich Observatory archives at Cambridge contain a few Kew solar photographs, including some from August and September 1859, it appears that none was taken on 26 March 1859.

9 See George C. Comstock, 'James Craig Watson, 1838–1880', *Biographical Memoirs of the National Academy*, III (Washington, DC, 1895), p. 54.

10 Ibid.

11 William Sheehan, 'Christian Heinrich Friedrich Peters, September 13, 1813–July 18, 1890', *Biographical Memoirs of the National Academy of Sciences*, LXXVI, p. 289.

12 Though Separation died not long after the eclipse, abandoned when the Union Pacific railroad shifted its track south, there are still a few remains, as I found on a visit in 2010, and one can make out the sand dune where Watson set up his telescope. Otherwise, the place is bleak and forlorn, surrounded by a rough alkali plain. It would be hard to imagine a more desolate scene.

13 For Watson's account, which is followed here, see J. C. Watson, 'On the Intra Mercurial Planets; From Letters to the Editors, Dated Ann Arbor, Sept. 3d, 5th and 17th, 1878', *American Journal of Science and Arts*, 3rd ser., XVI (1878), p. 310.

14 Christian Heinrich Peters, 'Some Critical Remarks on So-called Intramercurial Planet Observations', *Astronomische Nachrichten*, XCIV/2253 (1879), p. 321.

15 Agnes M. Clerke, *A Popular History of Astronomy during the Nineteenth Century* (London, 1908), p. 250.

16 Donald E. Osterbrock, 'Lick Observatory Eclipse Expeditions', *Astronomical Quarterly*, 3 (1980), p. 70.

17 E. S. Holden, 'Report on the Eclipse of May 6, 1883', *Memoirs of the National Academy of Sciences*, 2 (1884), p. 100.

18 William Wallace Campbell, 'The Crocker Eclipse Expedition of 1908 from the Lick Observatory, University of California', *Publications of the Astronomical Society of the Pacific*, XX (1908), p. 79.

19 Simon Newcomb, 'Discussion and Results of Observations on Transits of Mercury from 1677 to 1881', *Astronomical Papers of the American Ephemeris and Nautical Almanac*, 1 (1882), p. 367.

20 J. Laskar and M. Gastineau, 'Existence of Collisional Trajectories of Mercury, Mars and Venus with the Earth', *Nature*, CDLIX (11 June 2009), p. 817.

21 A. J. Steffl et al., 'A Search for Vulcanoids with the STEREO Heliospheric Imager', *Icarus*, CCXXIII/1 (March 2013), p. 48.

22 For a lucid popular account of these rather esoteric matters, see Konstantin Batygin, Gregory Laughlin and Alessandro Morbidelli, 'Born of Chaos', *Scientific American* (May 2016), p. 28.

23 Frederick Masset and Mark Snellgrove, 'Reversing Type II Migration: Resonance Trapping of a Lighter Giant Protoplanet', *Monthly Notices of the Royal Astronomical Society*, CCXX/4, pp. L55–L59.

24 Kevin J. Walsh et al., 'A Low Mass for Mars from Jupiter's Early Gas-driven Migration', *Nature*, CDLXXV (14 July 2011), p. 206.

25 Batygin et al., 'Born of Chaos', p. 32.

26 See Sean M. Mills et al., 'A Resonant Chain of Four Transiting, Sub-Neptune Planets', *Nature*, 11 May 2016, available at www.nature.com.

FURTHER READING

Antoniadi, E. M., *The Planet Mercury*, trans. P. Moore (Shaldon, 1974)

Balogh, A., L. Ksanfomality and R. von Steiger, eds, *Mercury* (New York, 2010)

Baum, R., and W. Sheehan, *In Search of Planet Vulcan: The Ghost in Newton's Clockwork Universe* (New York, 1997)

Clark, P. E., *Mercury's Interior, Surface, and Surrounding Environment: Latest Discoveries* (New York, 2015)

Grego, P., *Venus and Mercury, and How to Observe Them* (New York, 2008)

Roseveare, N. T., *Mercury's Perihelion from Le Verrier to Einstein* (Oxford, 1982)

Rothery, D. A., *Planet Mercury: From Pale Pink Dot to Dynamic World* (New York, 2015)

Sandner, W., *The Planet Mercury* (London, 1963)

Sheehan, W., *Worlds in the Sky: A History of Planetary Discovery, from Earliest Times through Voyager and Magellan* (Tucson, AZ, 1992)

Strom, R. G., *Mercury: The Elusive Planet* (Washington, DC, 1987)

—, and A. L. Sprague, *Exploring Mercury: The Iron Planet* (New York and Chichester, 2003)

Vilas, F., C. Chapman and M. S. Mathews, eds, *Mercury* (Tucson, AZ, 1989)

Westfall, J., and W. Sheehan, *Celestial Shadows: Eclipses, Transits, and Occultations* (New York, 2015)

Acknowledgements

I would like to acknowledge the contributions of many colleagues in making this book possible. When the idea of putting together a series of guides about the planets and other bodies of the Solar System was first considered, Thomas Hockey embraced it with enthusiasm. He has helped in innumerable ways from the beginning to the end of the project. Peter Morris read an early draft, and recommended its publication to Reaktion Books. My interest in Mercury has spanned decades, and at various points along the way I have received contributions from many colleagues (and fellow 'Hermophiles'): Richard Baum, Terrestrial Planets Section director of the British Astronomical Association (BAA) and an old friend, who took an interest in Mercury when it was largely seen as an unfruitful object of study; David Graham, former BAA Mercury Section director; Dieter Gerdes, who was a great help on Schröter; Luigi Prestinenza and Alessandro Manara on Schiaparelli (the latter of whom provided copies of all of Schiaparelli's Mercury drawings from the archives at Brera); John Boudreau, an exceptional CCD imager who joined the author's 'Schiaparelli project' and patiently accrued images of Mercury in the modern era corresponding to those in Schiaparelli's observing logbooks; Richard J. McKim and Patrick Moore on Antoniadi; Henri Camichel and Audouin Dollfus, who gave their insights into the Pic du Midi observations and generously provided copies of their observations and maps; and above all Dale Cruikshank and Clark R. Chapman who, though now distinguished planetary scientists, began as visual observers of the Moon and planets with small telescopes, and gave special attention to Mercury at a time when it was largely ignored.

Because of its position as the innermost planet, Mercury's story is tied in with the charming – and important – tale of the hypothetical planet Vulcan. I would like to thank Richard Baum, with whom I collaborated on the definitive book on Vulcan twenty years ago now, and who has remained the authority in the field; Kenneth Young, professor of physics at the Chinese University of Hong Kong, and his students, Sai Ming Chan and Xiaohe Zhou, who kindly recalculated the mass of an intra-Mercurial planet (or ring) needed to account for the anomalous precession of Mercury; James Lequeux, emeritus astronomer at the Paris Observatory, for his insights into Le Verrier's brilliant but controversial career; and Gregory Laughlin and Jacques Laskar for their helpful insights into some thorny problems of celestial mechanics.

Both David T. Blewett and Clark R. Chapman read an early draft of the entire manuscript in detail, and offered many useful suggestions. Any deficiencies that remain are, of course, my responsibility.

Photo Acknowledgements

The author and publishers wish to express their thanks to the below sources of illustrative material and/or permission to reproduce it. Some locations of artworks are also given below, in the interests of brevity:

Courtesy Leo Aerts: p. 32 (bottom); from E. M. Antoniadi, *La Planète Mercur* (Paris, 1930): pp. 12, 27, 72, 74; courtesy Arizona State University: p. 87; diagram Julian Baum: p. 134; courtesy Richard Baum: pp. 32 (top), 44 (bottom); images John Boudreau and William Sheehan p. 82 (both); courtesy Dale P. Cruikshank: p. 81; from W. F. Denning, *Telescopic Work for Starlight Evenings* (London, 1891): p. 43; courtesy Dr. Brett Denevi, JHAPL: p. 117; Dibner Library for the History of Science and Technology/Smithsonian Institution Libraries: p. 58; courtesy Audouin Dollfus and Clark R. Chapman: p. 83 (both); from J.L.E. Dreyer, *Tycho Brahe* (Edinburgh, 1890): p. 21 (left); courtesy Alan Dyer: p. 11 (top); courtesy Jo Edkins: p. 10; from Camille Flammarion, *Le Terres du Ciel* (Paris, 1884): pp. 28, 31; courtesy Mario Frassati: pp. 84, 85; from Pierre Gassendi, *Mercurius in sole visus* (Paris, 1656): p. 21 (right); courtesy Owen Gingerich: p. 15; Library of Polish Academy of Sciences, Gdańsk: p. 24; courtesy Lunar and Planetary Laboratory, University of Arizona: p. 90; © courtesy Alessandro Manara, Brera Observatory, Milan: pp. 56, 63; courtesy Alessandro Manara, Brera Observatory/John Boudreau: pp. 57 (both), 59 (both); from H. McEwen, 'The Markings of Mercury,' *Journal of the British Astronomical Association*, XLVI/10, (1936): p. 75; NASA: pp. 44 (top), 110; NASA /Applied Physics Laboratory (APL)/Carnegie Institution of Washington (CIW): p. 114 (top); NASA / Jet Propulsion Laboratory (JPL): pp. 91, 95, 97,99; NASA /John Hopkins University Applied Physics Laboratory (JHUAPL)/CIW: pp. 6, 96, 98, 100, 107, 108, 109, 112–13, 115, 114 (bottom), 116 (both), 118, 119 (both), 120, 124 (both), 125 (all), 126, 128; NASA/JHUAPL: p. 106; NASA/JHUAPL/CIW/National Astronomy and Ionosphere Center, Arecibo Observatory image: p. 127; courtesy NASA/JPL-Caltech/UCLA/ MPS/German Aerospace Center (DLR)/International Docking Adapter (IDA): p. 103; courtesy NASA/JPL/Northwestern University: p. 93; NASA/United States

Index

Page numbers in **_bold italics_** refer to illustrations

Abu-Nuwas (crater) 93
albedo 39, 119
    maps **83**
Alfonso x of Castile 17
Almagest (book by Ptolemy) 13
Angelico, Fra 132
Antoniadi, Eugène Michel 7, 10, 43,
    66–74, 80, 86, 100, 101
    chart of Mercury **74**
Apollonius of Perga 13
Arago, François 232
Arcetri Observatory 46
Arecibo Observatory 77
Aristarchus of Samos 12
Association of Lunar and Planetary
    Observers (ALPO) 80
asteroids 38, 103, 114, 120, 147, 170
Astronomische Nachrichten 143

Bach (crater) 93
Ball, Leo de 39
Bardou refractor, Juvisy **70**
Batygin, Konstantin 151
Beethoven (crater) 93
BepiColombo 8, 128
Bessel, Friedrich Wilhelm 37, 38
Bidault de l'Isle, Georges 66
Bi-ib-bou (Sumerian name of Mercury)
    10

Binder, Alan, 76
Birmingham, John 39
Borealis Planitia (Northern Plains) 103,
    115, 117
Boscovitch, Roger 48
Botticelli, Sandor 132
Boudreau, John 83
Brahe, Tycho 18, 19, **20**, 29–30
    Tychonic system **21**
Brera Observatory 43, 84
Brera Pinacoteca 53
Bridger, Jim 139
British Astronomical Association
    (BAA), 68, 106
Browning, John 39
Bunsen, Robert 133
Bureau des Longitudes 133

Caloris basin (Basin of Heat) **6**, 98, 99,
    99, 100, 103, **110**, 152
Caloris Montes 98
Camichel, Henri 74
Campbell, William Wallace 145
Caravaggio, Michelangelo 53
Carlini, Francesco 46
Carroll, Lewis 150
Cassini, Giovanni Domenico 30
Cayley plains 101
Ceres 152

Cervantes (crater) 93
Chao-Meng Fu (crater) 124
Chapman, Clark R. 76–7, 84, 103
Charge Couple Device (CCD) 33, 92
Charles II 24
Chopin (crater) 93
chromatic aberration 49
Cincinnati Observatory 132
Clarendon, Lord 48
Clerke, Agnes 143
Colombo, Giuseppe ('Bepi') 78
Columbus, Christopher 48
comets 25, 37, 38, 49, 89, 103, 114,
     120, 124, 128, 133, 138, 141, 142,
     143
Copernicus, Nicolaus 9, 15, 18, **19**, 29
Copley (crater on Mercury) 98
Correira, Alexandre C. M. 88
Côte d'Azur Observatory 151
Cruikshank, Dale P. 76, 80–81, **80**, 84,
     105
     observations composite **81**
cylindrical projections **82**

Darwin, Charles 133
     On the Origin of Species 133
Darwin, George Howard 58, **58**
Debussy (crater) 98, **109**, 111, **114**
Degas (crater) 42, 98, 101
Delambre, Jean-Baptiste-Joseph 25,
     131
De la Rue, Warren 39
Denning, William Frederick 41–5, **42**,
     52, 66, 92, 100
     Telescopic Work for Starlight Evenings
     42
     observation of Mercury **43**
Deslandres, Henri 68
Dollfus, Audouin 74, 80
Dollond, Peter 33
Donati, Giovanni Battista 48

Dostoyevsky (crater) 93
Dovo, Paolo 46
Draper, Henry 139, **140**
Draper, Mrs 139–40, **140**
Duccio (crater) **115**
Dyce, R. B. 78

Edison, Thomas 140
Einstein, Albert 146, **146**
Enterprise Rupes **116**
epicycle 14, **15**
Euclid 13
European Space Agency 128
Eurynome (asteroid) 138

Faye, Hervé 136
Flammarion, Camille 28, 68, 137
     depiction of Mercury **31**
     Juvisy chateau **69**
Flamsteed, John 31, 39
Fontenelle, Bernard le Bovier de 30
Fournier, Georges 66, 68
Fournier, Valentin 66
Frassati, Mario
     Mercury observations **84**, **85**

Galileo Galilei 22, 29
Gallet, 9
Ganymede 86
Garibaldi, Giuseppe 46
Gassendi, Pierre 22, 23, 31
     drawing of Mercury **21**
Gastineau, Mickaël 146
Gefken, Harm 35
General Relativity 2, 146
George III, King 35, 37
Gingerich, Owen 14, 40
Goldstone Solar System Radar 104
Goud 9
Goya (crater) 93
Graham, David 106, 109

Hall, Asaph 145
Halley, Edmond, 24, **25**, 131
Harding, Karl 36
Hellas 98
Herschel, William 33
Hevelius, Johannes 23, **23**, **24**
Hipparchus of Nicaea 13
Hiroshige (crater) 93
History of the Telescope (King) 35
Hokusai, Katsushika 114
Hokusai (crater) 114, **114**, **125**
Holden, Edward Singleton 144
Homer (crater) 93
Huggins, William 39
Hun Kal (crater) 99–100
Huygens, Christiaan 24, 30

Icarus 103
inferior planet **12**
International Astronomical Union
      (IAU) 93, 99

Jarry-Desloges, René 66
Juno 38
Jupiter 7, 31, 34, 35, 41, 46, 69, 86, 89,
      104, 149–52
      Great Red Spot 41

Kandinsky (crater) 124
Kästner, Abraham Gotthelf 33
Kepler, Johannes 15, 19, 20, **20**, 22, 23
King, Henry 35
Kirchoff, Gustav 133
Kuiper, Gerard P. 93
Kuiper (crater) 93, 98, **98**, **114**
Kuiper Belt 149

Lalande, Joseph-Jérôme Lefrancois de
      25, 131
Laplace, Pierre Simon de 150
Laskar, Jacques 88, 146

Laughlin, Gregory 151
Lescarbault, Edmond 136, 137
Le Verrier, Urbain Jean Joseph **130**,
      132–4, 136–7
      Théorie du mouvement de Mercure 132
Liais, Emmanuel 137
Lindenau, Bernhard August von 131, 132
Li Po (crater) 93
Lippi, Fra Lippo 132
Lowell, Percival 45, 49, 66, 69
Lyot, Bernard 74

Manara, Alessandro 84
Mare Orientale (lunar multi-ring basin)
      98, 117
Margherita, queen of Italy 53, 64
Mariner 4 80
Mariner 10 92–101, 104, 105, 108, 115,
      128, 147
      images captured by **93**, **94**, **97**, **99**
Mark Twain (crater) 93
Mars 8, 14, 15, 19, 27, 31, 35, 38, 43,
      44, **44**, 49, 51, 52, 54, 55, 69, 71,
      72, 74, 80, 83, 89, 95, 98, 101, 103,
      120, 145, 146, 147, 151
      canals of 44, 51
Masset, Frederic 152
Melville (crater) 93
Mendelssohn basin **121**
Mercury 11, **66**
      albedo 39
      apparent diameter 7
      atmosphere 31 39, 61, 62, 64–5,
      73–4, 86–7
      craters, rayed, seen from Earth 42,
      92–3, 100–101
      density 102
      diameter 86
      escape velocity 86
      'hot poles' 87, 98, 99
      inhabitants 30, 65

interior 101–2
iron core 8, 102, 128
librations 60, 61, 73
lobate scarps 101, 115–16
magnetic field 95, 101, 102, 127
and Moon 44
motions, as seen from Earth 12, 13–14
mountains 36
naked eye visibility 7, 9, 11
names 9, 10, 11
north pole 126, 127
orbit 20, 27–8, 27, 87, 103, 134, 137, 146, 147, 150
perihelion 134
phases 29
plains 117
planispheres 75
polar ice 124, 127
rotation period 8, 36, 37, 39, 41, 54, 65, 68, 70, 73, 76, 77–8, 79–85, 87
size of 28
south pole 128
southern horn, blunting of 36, 36, 115
spacecraft to 8, 92–3; see also Mariner 10
superior conjunction 72
tectonic activity 115–16
temperature 70, 89, 104, 124
trajectories to 90
transits 20–26, 22, 31–2, 32
'twilight zone' 61, 86
water 124
Mercury (god) 10
Mercury Dual Imaging System (MDIS) images by 115, 118, 119, 120
Merz refractor 48–9, 48, 50
MESSENGER (Mercury Surface, Space Environment, Geochemistry and

Ranging mission) 6, 8, 96, 98, 99, 104–6, 108, 108, 111, 114, 115, 116, 117–20, 118, 119, 120, 124, 124, 125, 126, 127–9
artist's conception of 106
colour base map 112–13
and General Relativity 146
launch 107
shaded relief map 122–3
the 'spider' 125
Meudon Observatory 69, 71
Michelangelo (crater) 93
Mitchel, Ormsby Macknight 132, 133
Moigno, Abbé François 136
Moon, the 7, 8, 11, 28, 29, 30, 33, 34, 38, 29, 44, 46, 51, 52, 53, 55, 56, 58, 60, 61, 64, 65, 70, 72, 77, 80 86, 89, 90, 92, 93, 95, 96, 98, 99, 101, 102, 103, 104, 105, 111, 114, 117, 118, 119, 120, 138, 139, 142, 145, 147
compared to Mercury 44
Mount Wilson Observatory 70
multi-ring basin 98–9
Murray, Bruce 91, 95–6

Nabou (Babylonian name for Mercury) 10
Napoleon 24, 37
Nasmyth, James 40
Near-Earth Objects (NEOs), 103, 147
Nervo Formation 98
Newcomb, Simon 140, 142, 145
Newton, Isaac 25
Nicholson, Seth B. 70

Odin Formation 99
Olbers, Heinrich 38

Pallas 38
Paris Observatory 132, 146

Peters, Christian Heinrich Friedrich 138–43
Petit, Edison 70
Pettengill, Gordon 78
phases of planets *30*
Pic du Midi Observatory 74, **76**
Pickering, William H. 144
Pliny the Elder 14
Pluto 149
Poggendorff's Jubelband 39
Prince, Charles Leeson 39
Principia (Newton) 25
Prokofiev (crater) 124
Proust (crater) 93
Ptolemy 13–18, **17**

Rachmaninoff (crater) **108**, 117, **118**
radio astronomy 77–8, 124
Raditladi basin **124**
Raphael (Raffaello Sanzio) 53
Renoir (crater) 93
Rodin (crater) 93
Royal Academy of the Lynxes 64
Rudaux, Lucien 66

Sagan, Carl **91**, 93
Scarlatti basin **125**
Scheiner, Christoph 22
Schiaparelli, Giovanni 7, 43, 44, 45–9, **47**, 51–62, **63**, 64–6, 69, 71–4, 80–85, 86, 93, 101
    little planisphere **64**
    logbook **56**, 57
    sketches **59**
    Sulla rotazione e sulla costituzione fisica del pianeta Mercurio 62
Schickard, Wilhelm 22
Schrader, Johann Gottlieb Friedrich 35
Schröter, Johann Hieronymus 31, 33–8, 39, 53
    drawing of Mercury *36*

Seeliger, Hugo von 145
Selenotopographische Fragmente (Selenotopographic Fragments) (Schröter) 34
Sella, Quintino 48
Shakerley, Jeremy 24
Shakespeare (crater) 93
Sharp, Abraham 31, 39
Snellgrove, Mark 150
Snorri (crater) 98
Sobkou (Eygptian name for Mercury) 10
Societas Lilitalica 35
Société Astronomique de France (SAF) 68
Solar Terrestrial Relations Observatory Heliospheric Imager 147
Solitudo Agriphontae 73
Solitudo Aphrodites 73
Solitudo Atlantis 72, 100
Solitudo Criphori 73, 100
Solitudo Hermae Trismigesti 73
Solitudo Horarus 100
Solitudo Lycaeus 100
Solitudo Martis 100
Solitudo Phoenicis 100
Sophocles (crater) 93
spin-orbit coupling 78–9, **79**, **87**, 88, 129
Stilbon (Greek name for Mercury) 11, 52
Strolling Astronomer (journal) 80
sunspots 22
Swift, Lewis 142, 144
synodic period 12
Syrtis Major 83

tasimeter 140
Terby, François, 59, 62
thermocouple 70
tidal friction, and rotation of Mercury 58, 61, 87

Tolstoy, Leo 57
    War and Peace 37
Tolstoy (crater) 93
total solar eclipse **134**
Tycho *see* Brahe, Tycho
Tycho (lunar crater) 111, 114

Umberto I, king of Italy 53, 64
Uranus 34, 133
U.S. Geological Survey 147

Van Allen radiation belt 95
van Eyck Formation 99
Vandamme (French general) 37
Van Gogh (crater) 93
Venus 10, **11**, 19, 20, 21, 27, 34, 40, 46,
    69, 78, 88. 89, 90, 92, 95, 101,
    105, 151
    transits of 21, 25
Very Large Array (VLA) 104
Vesta 38, **103**
Victoria Rupes **116**
Vittorio Emanuele II, king of Italy
    46
Vogel, Hermann 39
Vulcan 131–52
Vulcanoid Zone **147**
vulcanoids 148

Wagner (crater) 93
Walsh, Kevin J. 151
Watson, James Craig 138–43, **140**,
    144
Watt, James 17
Wells, H. G. 41
    War of the Worlds 41
Whitbread Engine **16**
Wren (crater) 93

Yeats (crater) 93
Yerkes Observatory 76

Zodiacal Light 145
Zola (crater) 93
Zöllner, Karl 39, 61
Zupus, Giovanni 29